Science and Medicine in France

Science and Medicine in France

The Emergence of Experimental Physiology, 1790–1855

John E. Lesch

Harvard University Press
Cambridge, Massachusetts
and London, England
1984

Library of Congress Cataloging in Publication Data

Lesch, John E., 1945–
  Science and medicine in France.

  Includes bibliographical references and index.
  1. Physiology, Experimental—France—History.
I. Title. [DNLM: 1. Physiology—History—France.
2. Pharmacology—History—France. QT 11 GF7 L6s]
QP21.L38 1984   612′.0072044   83-12749
ISBN 0-674-79400-1

To my parents

Acknowledgments

I am much indebted to colleagues for their support, criticisms, and suggestions. William Coleman, Roger Hahn, John L. Heilbron, Frederic L. Holmes, and Ralph Kellogg provided invaluable assistance, and I owe them special thanks. Among those who read versions of the manuscript in whole or in part are Gert Brieger, Gerald L. Geison, Lynn Hunt, James Kettner, and Karen Reeds. On research and technical matters I benefited from the advice of Toby Gelfand, Mirko D. Grmek, and Jeffrey Winer. My students Jane Phillips and David K. Robinson also contributed important comments.

It is a pleasure to record my particular debt to two people: Charles C. Gillispie, who first set this project in motion and has ever since helped sustain it as teacher and friend; and Paula Fass, who as colleague and critic made my work better, and as constant friend, companion, and mother of our daughter made it a joy.

Research for this book was supported in part by grants from the National Science Foundation, the National Endowment for the Humanities, and the University of California, Berkeley. Thanks are due the following institutions and their staffs, who facilitated the research or writing: in Paris, the Archives de l'Académie des Sciences, the Bibliothèque Nationale, the Bibliothèque de la Faculté de Médecine, the Bibliothèque de la Faculté de Pharmacie, the Musée d'Histoire de la Médecine, and the Centre Alexandre Koyré d'Histoire des Sciences; in Jerusalem, the library of the Hebrew University; in the United States, the libraries of the University of California, Berkeley, the University of Michigan, Princeton University, the New York Academy of Medicine, and the College of Physicians, Philadelphia, as well as the medical sciences library of the University of California, San Francisco, especially Nancy Zinn, the director of its history of medicine section. Crucial services were also provided by the staff of the Office for the History of Science and Technology, University of California, Berkeley, especially Jacqueline Craig and Amy Kivel; and by the staff of the History Department of the same institution. Florence Meyer produced a superb typescript, and Ellis Myers of Design Enterprises, Berkeley, provided the illustrations.

Finally, I would like to thank Aida Donald and Maria Ascher, who guided the book to publication with professionalism, efficiency, and care.

Contents

Introduction   *1*

1   A Science in the Making   *12*

2   Context for Change: The 1790s   *31*

3   Bichat's Two Physiologies   *50*

4   A New Generation and a New Program   *80*

5   The Experimentalist in Action   *99*

6   Pharmacists and Chemists   *125*

7   Experimental Pharmacology   *145*

8   Pathological Physiology   *166*

9   From Medicine to Biology   *197*

Notes   *227*

Index   *269*

Science and Medicine in France

Introduction

It is scarcely a century since the ideal of experimental control of the phenomena of life began to gain ascendency in the biological sciences. That ideal now dominates biology, despite divergences of specialty or theoretical orientation. It was not always so; and in the last decades of the nineteenth century and in the years after 1900, experimentalism had to make its way against the morphological and descriptive approaches that prevailed in many fields. Partly with the aim of ending the confusion and uncertainty of evolution theory, a generation of embryologists and students of heredity turned to the laboratory and experimental garden. Although the emergence of classical genetics after 1910 was the most striking vindication of their efforts, their more fundamental legacy was the consolidation of the ideal and methods of experimentalism in biology.

If dissatisfaction with contemporary evolution theory provided much of the motivation for the change, its positive exemplar came from physiology. By the 1870s physiology was firmly consolidated as a discipline, with specialized university chairs and laboratories in existence in France, Germany, England, and the United States.

Career training was available, especially in the lavishly endowed institutes of the German universities, such as Carl Ludwig's at Leipzig. More to the point, physiology had long since established its identity as an experimental science. The classic articulation of that identity was published by Claude Bernard in 1865, the same year that Ludwig's Leipzig institute opened, as *Introduction à l'étude de la médecine expérimentale*. Bernard emphasized experimental determinism, the invariable replication of the same phenomena under the same experimentally controlled physical and chemical conditions, as the goal and the method of physiology: the goal, because science could expect no more—and settle for no less—than the experimental or mathematical determination of the relations among phenomena; the method, because the means to that end in the sciences of life could only be experiments on living plants and animals, conceived, carried out, and interpreted with the aid of the adjunct sciences of chemistry and physics. Bernard stated unequivocally that not only physiology and the medical sciences but also botany and zoology would realize their full potential as sciences only by adopting and practicing these principles. In the last decades of his life he began to cultivate general physiology, defined as the study of phenomena common to plants and animals. To the student of evolution, embryology, or heredity casting about for a path out of mere description or the uncertainties of speculation, the undeniable successes of physiology made it an obvious object of emulation.

The expansion of the ideal and practice of experimentalism in biology after about 1870 has yet to be given definitive treatment by historians. Whatever the extent of its role in that process turns out to be, it is certain that physiology was the first of the life sciences to commit itself fully to the experimental ideal in the sense given it by Bernard. It did so in a process that began long before Bernard's time, though he is rightly regarded as the outstanding representative of the trend and its most articulate spokesman. The pivotal movement that brought experimentalism from the periphery to the center of physiological thought and practice occurred in France in the last decade of the eighteenth century and the early decades of the nineteenth. Though at first confined to a relative handful of adepts, experimental physiology gradually displaced alternative approaches. By the 1840s it was beginning to take hold in the German universities, and by the 1870s in those of Britain and the United States.

In view of the central place that experimentalism came to occupy in modern biology, and the leading role played by physiology in its assimilation by the life sciences, its early origins in French physiology are of special interest. This study seeks to answer some basic questions raised by those origins. Why did experimental physiology appear in France around 1800, and why did it take the form it did? How were the timing of its appearance and the subsequent evolution of its content conditioned by the Paris scientific and medical milieu in all its aspects?

Pursuit of answers to such questions requires an approach that is neither internalist nor externalist, though it shares features of both. Science is, first and foremost, knowledge about nature, whether that knowledge aims for theoretical understanding or practical control. This much admitted, the methods and techniques by which knowledge is acquired, and the facts or concepts that constitute its content, must be objects of the highest interest to historians. The history of science cannot, however, confine itself to study of ideas and techniques, their form, content, and filiation. Obviously, science has been more than that. It has had numerous ramifications in other areas of thought, occasionally provoking convulsions in them. It has contributed to the esoteric body of knowledge of several of the learned professions. It has given rise to learned societies. It has informed experts in service to the state. It has helped motivate and implement exploration of the globe. Especially in the last century, it has provided the basis for whole industries and through them has transformed economic and social life and the nature of war.

What remains to be asked is how, and to what extent, the content of science—its self-definition, concepts, methods, techniques, and instruments—has been shaped by the circumstances in which it has arisen and has been cultivated. Both internalist and externalist rightly concede at least some degree of autonomy to science with respect to other spheres of human activity. The challenge facing historians is to show how the sciences, while retaining this necessary degree of autonomy, have also been creatures of time and place.

Though the legitimacy and urgency of the problem have been

widely recognized, its translation into finished historical studies that display answers in concrete form has proved difficult. Part of the difficulty stems from insufficient attention to the immediate environment of science: the ideas and institutions that provide the work and career setting of the scientist, in contrast to the larger social or political scene. This immediate environment is shaped by the larger one. But often events in society at large (economic growth, war, revolution, for example) affect the scientific work of the individual only indirectly, through their effects on institutions important for his (or her) training, professional identity, employment, or learned associations.

This book focuses on the interaction between the scientist and his immediately experienced environment, and especially on the ways in which this environment contributes to the formation of the individual's science. I have sought to adhere to rules of method that seem to me fundamental for the history of science. Institutional or intellectual context and the conceptual and technical content of science should be studied together. In the case of the appearance of experimental physiology in France, neither timing nor content can be understood apart from such an integrated approach. Conclusions drawn regarding the connection of context and content can be valid only if the integrity of the conceptual and technical dimensions of the science, including its unspoken or craft aspects, is preserved. Programmatic or philosophical statements by past scientists should be taken seriously, but not necessarily at face value, and on no account should they be allowed to represent or substitute for the scientist's concrete research.

My aim has not been to present a comprehensive history of physiology in this period in France. I have concentrated on the scientific careers of three men—Xavier Bichat, François Magendie, and Claude Bernard—who were indisputably the most important physiologists of their time and whose careers and work were in many respects representative of their science. Even for them I have been selective, emphasizing aspects of their scientific careers that have received insufficient attention from historians and that illuminate the shaping of experimental physiology by its environment over more than a half century of its formative development.

Experimental physiology took form in a period of French ascendency in medicine and science. The investigations of Pierre

Simon Laplace in celestial mechanics and probability, of Joseph Gay-Lussac in chemistry, and of Georges Cuvier in comparative anatomy were only the most outstanding examples of a wide range of scientific achievement centered in the first third of the century. The Paris clinical school that emerged in the years of the Revolution, First Empire, and Restoration opened a new era in medicine. The unification of surgery and internal medicine, the centering of practice and research in the hospitals, the invention and exploitation of new methods of physical diagnosis, and the cultivation of pathological anatomy in all its aspects won the admiration and emulation of Europeans and Americans for half a century or more. The timing of the appearance of experimental physiology, its form and subsequent evolution, are explicable only within the special environment created by these realities.

The imprint of the medical milieu, though deep, was in some respects negative. As Erwin Ackerknecht has emphasized, the tendency to exclusive concentration on clinical observation, pathological anatomy, and statistics that marked Paris hospital medicine could foster indifference to the experimental or laboratory sciences. Therapeutic skepticism, with its passive, expectant posture in the face of disease, might have the same effect. The de facto exclusion of many chemists, physiologists, and microscopists from positions in the Paris Faculty of Medicine over several decades appears to indicate long-standing attitudes unfavorable to the role of the sciences within medicine. Magendie's frequent complaints about the excessive empiricism and hostility to experimental physiology voiced by a segment of the Paris medical community provides further evidence to the same effect. Bernard's *Introduction à l'étude de la médecine expérimentale* is in one sense an extended critique of the limitations of clinical medicine vis-à-vis his science.

To a surprising degree, however, the influence of Paris medicine on experimental physiology was fundamentally constructive. The surgical training and practice of several of the early experimental physiologists, including Bichat, Guillaume Dupuytren, and Magendie, supplied them with institutional roles and material support. Professional work led Dupuytren away from research; but for those who continued, a surgical background provided attitudes, skills, and experience important to the new science. The activist impulse to intervene in the organism—an impulse characteristic of surgeons—helped motivate animal experiment, and the surgeon's operative facility and practical knowledge of gross anatomy supplied the foundations of technique. From clinical experience came suggestions of physiological problems to be investigated, a field of observation that complemented animal experiment, and the routine of autopsy associated with pathological anatomy. Where it did not dissolve theoretical interest altogether, the surgeon's skeptical reserve helped provide a needed corrective to eighteenth-century styles of hypothetical explanation in physiology. The depth of the surgical current in French physiology is still evident in Claude Bernard, who did not practice surgery but whose language, operative skill, instruments, routine practice of autopsy, and authorship of texts on operative physiology and medicine all carry forward the surgical tradition within his science.

The reformed medical schools that emerged from the convulsions of the 1790s gave to a new generation of French physicians the sort of training and experience previously characteristic only of surgical education. An emphasis on normal and pathological anatomy, practical exercises for students at all levels, and clinical instruction centered in the hospitals became the norm. At the revived Ecole Pratique the best students could and did try their hand at research. Among the early graduating classes were the first experimental physiologists of the new century: Pierre Nysten, Guillaume Dupuytren, Julien Legallois, and Magendie.

One source of the new medical education was the program and activity of the Société Royale de Médecine. Founded in the 1770s as part of a government effort to control epizootic disease, the Société expanded its efforts and official responsibilities to include public health. In contrast to the largely tradition-bound faculties of medicine, the Société vigorously promoted a unified conception of the healing arts and their association with the modern sciences. Among the latter physiology, in a form less hypothetical and more closely linked to anatomy, was to occupy an important place.

The Société Royale de Médecine embodied one constructive response to a profound sense of dissatisfaction with the theoretical validity and practical efficacy of internal medicine that developed in the last decades of the Old Regime and that became especially acute in the Paris scientific and medical community of the 1790s. Another form of response lay in the effort to reconstitute the scientific foundations of the constituent fields of internal medicine, especially physiology and materia medica, or drug therapy. Both

Bichat and Magendie simultaneously shared in the prevailing mood of skepticism and attempted to transcend it through positive efforts of reconstruction.

Some of the resources for that effort were provided by the newly professionalized healing arts of pharmacy and veterinary medicine. Like surgery, they had begun to win full professional status in the eighteenth century, and as part of that process had associated themselves with research and teaching of the modern sciences. Both professions supplied personnel, locales, materials, and particular results that either contributed directly to the substance of experimental physiology or provided it with research tools. Pierre Flandrin's studies of venous absorption, Magendie and Dupuytren's experimental work at Alfort, and the isolation of active principles of medicines by Joseph Pelletier, Joseph Caventou, and other pharmacists exemplify the process.

The existence of professional schools of pharmacy and veterinary medicine reminds us that not all instruction in medicine and related sciences took place within the Faculty of Medicine. The institution of the private course, long an important part of medical teaching under the Old Regime, continued to thrive in the nineteenth century. For innovative investigators like Bichat and Magendie, and for new fields like experimental physiology or pharmacology, the private course could and did provide a context for the development and diffusion of research results not dependent upon the official chairs of the Faculty.

Nor did the physiologist have to depend on a Faculty chair for a livelihood. In addition to private practice and the fees collected for private courses, he might also obtain a position on the staff of a Paris hospital. A hospital position provided not only a living and prestige but also continuous experience in the clinic and autopsy room that could be a rich source of physiological problems and observations. Bichat's surgical practice at the Hôtel-Dieu was an important source for his experimental physiology. Magendie's increasing association with Paris hospitals after 1818 encouraged and supplied essential materials for a program of pathological physiology that complemented and helped advance experimental physiology. Both in name and substance, the program was indebted to the pathological anatomy that was a more visible and dominant part of the Paris medical scene after 1815 and that in other respects is often seen as in opposition to the basic sciences. The wards of the Hôtel-Dieu and other hospitals also served as testing grounds for the therapeutic efficacy of medicines, in particular the new substances supplied by chemistry.

Magendie's investigations in experimental physiology and pharmacology and in pathological physiology were supported and assisted by many Paris clinicians and practitioners, and not only in their directly therapeutic or practical aspects. The pages of Magendie's *Journal* of the 1820s and of the later editions of his *Formulaire* in the 1830s provide ample evidence of a network of cooperation that involved many members of the medical as well as of the scientific communities in Magendie's programs.

The existence of a scientifically interested and committed segment of the Paris medical community was finally given institutional expression in the founding of the Société de Biologie in 1849. Among the objectives of the new society was the cultivation of a science of life that would not be preoccupied with immediate practical applications or with a narrow concentration on pathological anatomy. In its stated program and in the realities of its early membership and publications, however, the Société was deeply marked by its origins in the Paris medical milieu of the preceding decades. Only gradually would the ideal of biology as a general science of life begin to be realized.

The beginnings, but only the beginnings, of the transition from medicine to biology are visible in the early career of one of the first vice-presidents of the Société de Biologie, Claude Bernard. In the first—and most creative—decade of his career, Bernard retained the emphasis on human physiology and the mix of experimental and pathological approaches that he had inherited from his teacher, Magendie. Only in his greater subordination of pathology to physiology and in his abandonment of medical practice is movement away from a medical orientation evident. It was only in the years after 1850 that he began to broaden his interests beyond higher mammalian function to general physiology, or study of the phenomena of life common to plants and animals.

In its personnel, sources of support, and content, therefore, the experimental physiology of the first half of the century was shaped in many ways by the Paris medical environment. Yet it was as a science that Bichat, Magendie, and others sought self-definition and recognition for the field, and neither the origins of that ambition nor its gradual realization is explicable apart from the relations of physiologists to the Paris scientific community. The extraordinary strength of contemporary French science provided physiologists with exemplars of what a science was, a strong sense of the norms and standards of research, and a source of substantive knowledge and technique that was of instrumental value to the new field.

A central role in bringing these influences to physiology was played by the First Class of the Institute, or Academy of Science. A crucial step was taken when separate sections for the healing arts were included in the First Class on its organization in 1795. Henceforth medical men might aspire to membership in this prestigious body, and once there would mix regularly with the elite of French science. The younger generation responded, and experimental physiologists such as Bichat, Nysten, Dupuytren, Legallois, and Magendie made recognition by the First Class a career goal. In doing so they strove to adapt their programs and practices to what they perceived to be the standards of the established fields of astronomy, physics, chemistry, and comparative anatomy. The leading members of the First Class, including Laplace, Cuvier, and Claude Berthollet, in turn encouraged and attempted to channel the efforts of the younger men. Above all, they sought to reinforce the development of a rigorous experimentalism in physiology. They also increasingly urged the enrichment of the field by physical, chemical, and comparative approaches. Creation of the Montyon prize in experimental physiology in 1818 signaled the Institute's formal recognition of the field as a science, and Magendie's election to membership three years later further reinforced the trend.

Comparison of the context of thought and institutions in German-speaking lands and in Britain offers some perspective on French leadership in the introduction of experimental physiology. German schools and universities placed little emphasis on empirical or mathematical sciences before the 1820s. Romantic *Naturphilosophie*, until its demise in the 1820s and 1830s, provided an idealistic alternative. More important than *Naturphilosophie* for German life scientists was a teleomechanist orientation, derived ultimately from Kant and exemplified for physiology in the work of Johannes Müller at Berlin. The holism of the teleomechanist program favored modes of investigation that allowed subordination of the parts to the whole. Study of morphology and ontogenetic development took priority over vivisection, which was seen as disruptive of organic relationships. The separation of the theoretical and practical in German medical education, and the early availability of full-time university positions for physiologists, tended to remove from German physiology the clinical and pathological interests and surgical orientation that were characteristic of French physiologists and that contributed to their experimentalism.[1]

A few British surgeons and physicians made promising beginnings in experimental physiology in the crucial formative decades between 1790 and 1820. By the 1830s, however, a variety of institutional, intellectual, and social factors had combined to prevent continuity of research efforts and the clear self-definition and institutionalization of the field. Chemistry, geology, and the natural-history disciplines absorbed much scientific talent through the first half of the century. The Industrial Revolution posed problems of public health for which physiology could not offer or even promise solutions. Deeply held utilitarian attitudes and, in the institutional sphere, traditional reliance on private initiative and funding, converged to preclude strong moral or material support for the pursuit of knowledge for its own sake. Natural theology, especially strong in the Anglo-American cultural area, stimulated and legitimized research, but also tended to subordinate scientific to moral or theological ends. Antivivisection sentiment discouraged adoption of animal experiment. So too did physicians' aversion to manual procedures, reinforced by a persisting and invidious distinction between physicians and surgeons in both education and social status. British medical schools emphasized teaching and practice at the expense of research.[2] The British pattern was replicated with variations in the United States before the Civil War.[3]

These comparisons suggest that what distinguished the French case was the concurrent presence of a vigorous hospital medicine and a strong, institutionalized commitment to basic research. German physiologists assimilated the profound theoretical orientation of their academic environments but were relatively insulated from the stimulus of clinical medicine. British and American physiologists enjoyed much of the practical training and experience of their French counterparts but were thwarted in their scientific endeavors by the relatively low levels of material and moral support for research provided by their societies. Only in France did physiologists trained in and practicing the new hospital medicine find inspiration, direction, and support for their scientific aspirations in a firmly institutionalized research tradition. In this environment the decisive steps were taken that brought physiology to its identity as an experimental science, and so ultimately to its role as principal exemplar for the increasing experimentalism of modern biology.

The time has passed when the teaching of physiology was composed of hypothetical explanations and in which books were filled with purely imaginary systems. . . . How different is the physiology of our day from that of the last century! By what precious and unexpected discoveries has it been enriched in the last twenty years! And it is above all to the French that this surprising progress is due.
—*Philippe Pinel and Pierre-François Percy* (May 1812)

Physiology was then being born. . . . Still not introduced into the former Academy of Science, it was omitted in the formation of the Institute . . . These brilliant acquisitions of the human mind were not foreseen.
—*Etienne Geoffroy Saint-Hilaire* (July 1821)

1 A Science in the Making

So low was the profile of physiology as a science in the late eighteenth century that when, in 1795, the elite of the French scientific world gathered to constitute the First Class of the National Institute of France, no section was designated for the subject.[1] Looking back twenty or thirty years later, Pinel, Percy, and Geoffroy Saint-Hilaire could tell their assembled colleagues in the same institution that physiology as a science was virtually a product of French efforts during those two or three decades, and no dissenting voices would be heard.

No doubt in making these claims the three savants were suffering from a case of historical myopia complicated by national sentiment, with a result bordering on caricature. In one sense the study of human and animal function, or physiology, was as old as medicine itself, certainly as old as the Hippocratic and Galenic writings with which most nineteenth-century physicians were familiar. Neither could they easily overlook the moderns, from Andreas Vesalius through William Harvey to Albrecht Haller. Whether or not the Institute had officially recognized the subject, its prior history should not have been dismissed.

In another sense, however, they and the Institute were right. No one then or now could reasonably describe the physiology of the 1790s as a coherent scientific field. What would later be called physiology was cultivated under different rubrics, by different schools with divergent theoretical and methodological orientations. Experiment, in the sense of surgical intervention to control bodily phenomena, still competed with other research methods for the allegiance of physiologists. Most prominent among the alternatives were classification of body tissues into categories according to their possession of one or more vital properties (variously defined), observation of the living organism in health and disease, and human or comparative anatomy. A savant of the 1790s might, but need not, make use of animal experiment, and even those who did so were likely to combine it with one or more other approaches. Institutionally as well as intellectually, physiology was hardly distinct from medicine. There were no facilities for specialized training or research, no specialized publications.

Pinel and Percy were also correct in their assertion that important changes in the content and status of physiology were initiated by French savants. Between 1790 and 1821 physiology in France achieved self-definition as a science, a degree of autonomy from medicine, and an official recognition from the scientific world— and these to an extent that existed nowhere else. Encouraged by leaders of the First Class of the Institute, members of a new generation of the Paris medical community undertook deliberate programs to construct a science of physiology. In their concrete research, animal experiment of a marked analytical and surgical quality occupied a central place. The physiology that took form was one closely linked with anatomy, in which the primary goal was to elucidate functional interdependences through direct surgical intervention in the processes of the living organism. It was analytical since each investigation aimed to isolate the structural elements on which a given function depended, and because the results of each investigation could stand alone, without being referred to some more general system of physiological explanation. Physiologists increasingly sought to make use of chemistry and physics as adjunct or tool sciences. Underlying the endeavor was a commitment to what Claude Bernard would later assert as the ultimate goal of experimental physiology: the operative control of the phenomena of the living organism. These traits, which characterized

French physiology by the 1820s, mark it as the direct ancestor of the mature laboratory science of the later nineteenth century, and of our own time.

Models for a physiology of experimental control had long been available in some of the work of Galen and a succession of moderns from the Paduan school through Harvey to Haller and Lazzaro Spallanzani. Before 1790, however, none of these models of experimentalism had succeeded in displacing other approaches to the science, and none had given rise to a continuous tradition of research. They could become meaningful for the emergence of experimental physiology as a scientific field only when an appropriate context was available—a context that would lead savants to conceive of that science as distinct from others, would give continuous support to their endeavor to construct it, and would encourage their primary reliance on animal experiment and the ancillary use of chemistry and physics.

It is no accident that the appearance of this context coincided in time and place with the French Revolution and its aftermath. The upheaval and transformation of French life from Revolution through First Empire to Restoration strongly marked the educational and professional institutions in which physiology and its adjunct sciences were cultivated. After centuries of separation, medicine and surgery were reunited, ending the invidious social distinction between the two professions and with it the stigma associated with manual operations. One consequence was a transformed medical education, with far greater emphasis on practical exercises, clinical experience in hospitals, and anatomy. Eighteenth-century trends toward the professionalization of pharmacy and veterinary medicine, which included increasing association between their personnel and the modern sciences, were continued and consolidated. Scientifically minded practitioners of the healing arts gained better access to the highest levels of French (and European) science when sections were designated for them within the First Class of the Institute on its organization in 1795.[2]

These changes contributed to the creation of an environment more favorable to the cultivation of experimental physiology than any that had existed before 1790 or that did exist in the contemporary scientific nations of Europe. The results are evident not only in the content of the physiology produced by French savants but in a number of visible tokens of the emerging field. Specialized textbooks of experimental physiology appeared in France in the 1810s, a specialized journal in 1821. The First Class of the Institute used the funds of a private donor to found in 1818 an annual prize in experimental physiology, and in 1821 elected the leading practitioner of the subject, François Magendie, to membership. Magendie's election occasioned debates on the status of physiology within the Institute that came near to winning the field a separate section. That did not happen, but the First Class did contribute to the coalescing identity of the field by its consistent use of the terms "physiology" and "physiologist" in their modern senses.

Consolidation of the name was important because it signaled the approaching end of nearly three centuries of fundamental disagreements among savants regarding the content and independent existence of the science. Physiology shared with physics a root in the Greek word for nature (*physis*); and the use of "physiology" to mean the study of nature persisted at least until the end of the eighteenth century.[3] By that time, however, this usage had been largely displaced by a narrower one, for which physiology was the study of the functioning of living things. The change was gradual, and as it occurred the study of living functions continued to be called by other names. Jean Fernel, writing a medical textbook in the mid-sixteenth century, was probably the first to use "physiology" in a sense approaching the modern one, but his practice was not generally adopted.[4] The Paduan school did not use the term. Nor did William Harvey, who called his work on the heart and blood "anatomical exercises." Iatrochemists and iatrophysicists, the latter including Descartes and Newton, wrote on physiology but seldom called it that. "The animal economy" was a common alternative, one that persisted in English-speaking lands into the nineteenth century. Hermann Boerhaave and many university teachers of the eighteenth and even nineteenth century used "institutes of medicine."[5] The physiological writers of the medical school of Montpellier traversed the whole of the eighteenth century without calling themselves physiologists or their work physiology. Something of a turning point occurred in the 1740s and 1750s when Albrecht Haller used "physiology" in the titles of his authoritative and widely influential textbooks. In 1755 the Paris Collège de Chirurgie changed the name of its course "Principles of Surgery" to "Physiology," and one of its professors translated Haller's *First Lines of Physiology* into French.[6] In 1795 the German physician Johann Christian Reil founded the *Archiv für Physiologie*, the first journal to incorporate the term into its title.[7] By the 1790s the name physiology had a wider currency than it did in 1700, but was still not generally employed. Its ascendancy in scientific and medical language would be won only in the course of the changes that were to take place over the next three decades in France.

Inconstancy of name reflected a diversity in programs, concepts, and methods that persisted throughout the eighteenth century. By around 1700 the study of human and animal function, or physiology, was pursued within a variety of conceptual frameworks and with a variety of techniques. The anatomical and experimental tradition that had found its greatest exemplar in Harvey was continued by Marcello Malpighi, whose microscopical studies confirmed the existence of capillaries; by Richard Lower and the Oxford school, who extended Harvey's studies on heart and blood function to include study of respiration; and by others.[8] The iatrochemical trend initiated by Paracelsus had been continued in the seventeenth century by Jan van Helmont and Franz de la Boë (Sylvius), and still found adherents in the beginning of the eighteenth century. The mechanistic approach to physiology continued to enjoy great strength, particularly in England in the ranks of the Royal College of Physicians, and in Italy, where the attempts of Giovanni Borelli and Georgio Baglivi to offer mechanical explanations of particular body functions reflected and reinforced a general consensus in favor of mechanism.[9] Robert Boyle's adaptation of the mechanical philosophy to chemical phenomena met with some success, finding application to physiological problems in the Oxford school and winning the adherence of Isaac Newton.

To some extent the various seventeenth-century trends were mutually contradictory: Borelli advanced his mechanistic explanations in place of chemical ideas; those immersed in the anatomical-experimental approach, such as Malpighi, tended to avoid deductive systems, whether chemical or mechanical. To some extent, these trends were overlapping: Franz de la Boë advanced chemical explanations of physiological phenomena while considering himself a mechanist; Boyle joined the mechanistic and chemical approaches by mechanizing chemistry; and the Oxford school joined the chemistry so modified to its anatomical-experimental investigations. In many medical faculties the old Galenic humoral physiology retained a strong hold. Harvey's discovery alone, and its extension in the researches of Malpighi, the Oxford school, and others, had greatly expanded the factual content of physiology and demonstrated the power of the anatomical-experimental approach. Iatrochemical and mechanistic ideas, though often overly deductive and simplistic, had enriched the theoretical options of physiology. Perhaps more important, they had created an attitude of openness toward the utilization of discoveries in physics and chemistry. Yet the general situation around 1700 was characterized by diversity: no single paradigm, or coherent body of concepts and techniques, had succeeded in winning general acceptance.

The obverse of this diversity in concepts and methods was the lack of distinct institutional status for physiology and the consequent absence of the continuity and stability such status brings. As long as physiology remained amorphous in self-definition, concepts, and methods, physiologists could not expect—and might not even think to seek—institutional support and recognition for their subject as a distinct field. And without separate institutional status, the field would be unlikely to achieve stability and continuity in its identity and content.

The strength and limitation of physiology in the eighteenth century derived from its integration with medicine. With few exceptions physiologists were trained as physicians or surgeons, and both their professional identities and their incomes came from association with one of the healing arts. Many taught in medical or surgical schools, where physiology found a place in the curriculum as "institutes of medicine" or "principles of surgery." Others drew a living from practice of their profession. For most, physiology—under whatever name—was a part-time activity, often without direct remuneration. They published their findings in medical or surgical journals or in monographs directed at least in part to the medical community. The leading scientific institutions of Europe, such as the Paris Academy of Science, also published some of their work, but did not recognize physiology as a distinct field.

Integration with the healing arts had consequences for the content of physiology that are evident, with local variations, in all the scientifically active areas of Europe. Concentration on the human organism tended to displace comparative studies of animal or plant function. Physicians gave special attention to the integrating, coordinating phenomena evident in medical practice and sometimes touted observation of the whole, living human in health and disease as the method of choice. Concern that knowledge inform therapy led to occasional subordination of scientific to utilitarian goals in research.

Some eighteenth-century physiologists fell entirely outside the medical pattern. They were not trained in medicine or another healing art, nor were their means of support associated with medicine. A few such individuals appeared in all the scientifically active areas of Europe, and together they formed a distinct species. Most often, like the medical people, they did not use the term "physiology." Unlike most of their medical counterparts, their interests tended to be more broadly biological, in the sense that they did not restrict their attention to human or analogous vertebrate function, but often studied plants or invertebrates. Some made use of experiment. Like the medical men, they usually influenced succeeding generations through teaching or publication, but did not establish institutionalized research traditions. Outstanding specimens are Abraham Trembley in Switzerland and Alexander von Humboldt in the German-speaking lands, Felice Fontana and Lazzaro Spallanzani in Italy, Stephen Hales in England, and René Antoine de Réaumur in France.[10]

This lack of differentiated institutional status and integration with the healing arts had important bearings on the status of experimentalism. Individual savants cultivated an experimental approach, sometimes with great success. Without formal mechanisms for recruiting, training, and employing successors, however, they had no way to assure the continuity of their own notion of the science or the continuity of methodological commitments. Though the results of their efforts usually reached the international scientific community through publications, they were unlikely to lead to the establishment of continuous lines of experimental research. Consolidation of any single methodology, including experimentalism, was inhibited by the absence of stable institutional forms: it was every man—or school—for himself. Physicians and surgeons—the majority of physiologists—were subject to the temptation of alternative methods, such as observation of the healthy or sick person, or reasoning from anatomical structure to function. In order to do experimental work at all, physicians had to overcome the stigma associated with manual operations, which had long been regarded as the province of the socially inferior surgeon. Institutional integration of physiology with the sciences, had it occurred, might have given experimentalism a more prominent and stable place. In its absence, the idea of what constituted a science of physiology, and with it the role of experimentalism, was bound to fluctuate.

The German-language areas of Europe provide instructive instances of the general pattern. The century opened with an impressive effort by Friedrich Hoffmann, professor of medicine at the newly founded University of Halle, to achieve a didactic synthesis of seventeenth-century trends in medical theory. No specialist, Hoffmann lectured on chemistry, physics, anatomy, surgery, and clinical medicine; at the same time he maintained a successful private practice that included three years as personal physician to Frederick I. His physiology, published as *Medicina rationalis systematica* (1718–1740), incorporated classical Galenic teaching, Cartesian iatromechanism, Boyle's chemistry, and the most recent anatomy and microscopy. The eclectic, discursive result favored mechanistic explanation, and Hoffmann's popularity as a teacher helped reinforce this already prevalent tendency in the medical thought of the early decades of the century.[11]

The first serious challenge to the uncritical consensus supporting iatromechanism came from a man Hoffmann had recruited to be professor of medicine at Halle, Georg Ernst Stahl. Better known as one of the authors of the phlogiston theory in chemistry, Stahl also practiced medicine, and after 1715 served as personal physician to Frederick William I. His attack on iatromechanism reveals the extent to which eighteenth-century physiology, still firmly embedded in medical theory, could be dependent on philosophical commitments and rational discourse, with profound theoretical divergences as the result.[12]

Stahl's animism was to have important long-term effects on eighteenth-century physiological thought, especially through the writers of the medical school of Montpellier.[13] The dominant trend of the first half of the century, however, continued to be an eclectic mechanism in the style of Friedrich Hoffmann, and of Hoffmann's still more influential Dutch counterpart, Hermann Boerhaave. As professor of botany, chemistry, and theoretical and practical medicine at the University of Leiden, Boerhaave was well placed temperamentally and professionally to produce a didactic synthesis of classical medicine with the most recently acquired knowledge in the sciences. His physiology, best represented in the *Institutiones medicae* of 1708, enjoyed a great success, and he remained a standard authority on medical theory through the mid-century and beyond.[14]

To the extent that anyone may be said to have displaced Boerhaave as an authority on physiology, that role fell to one of his Swiss students, Albrecht Haller. Taking his M.D. at Leiden in 1727, Haller pursued a varied career as physician; private teacher; professor of anatomy, surgery, and medicine at the University of Göttingen; and official in the municipal government of Bern.[15]

Haller's physiological writings embodied at least seven distinct research approaches. In his work on sensibility and irritability, he mapped the presence or absence of those properties over the entire body. The means was animal experiment and the definitions and criteria employed were behavioral, but the result was essentially classificatory: some parts (nerves) possessed sensibility, some (muscles) possessed irritability, the rest neither. In other research he took as the unit of study some function such as circulation, which was then analyzed by surgical intervention and manipulation of its component parts, or what Haller termed "anatomical experiments."[16] Both sorts of experiment relied on descriptive anatomy at the gross and microscopic levels, and in some of the physiological writings such description stood out as a third approach.[17] Comparative anatomical studies of animals supplemented those on humans. Studies of monsters and deformities were pursued to elicit clues to normal function. Following Boerhaave and the iatromechanical tradition, Haller did not remain at the phenomenal level but elaborated a speculative micromechanical physiology positing invisible mechanisms to account for macroscopic phenomena.[18] In his last years, when illness made it possible, he conducted experiments on himself to assess the effects of opium on the human body. Though used by Haller on occasion, the tag *anatomia animata* by itself fails to do justice to the diversity of his approach to physiology.[19]

Haller's efforts in physiology were also pedagogical and synthetic. His *Primae lineae physiologiae* of 1747, intended as an elementary introduction for students and physicians, was perhaps the first textbook of physiology. Not content with this beginning, he went on to produce a massive survey of contemporary knowledge in all fields of physiology, including historical summaries of the development of each area. This appeared in eight volumes from 1757 to 1766, as the *Elementa physiologiae corporis humani*. The magnitude of Haller's achievement was soon recognized in medical circles, and from the 1760s he began to displace Boerhaave as the standard modern authority in physiology.[20]

The careers of Hoffmann, Stahl, and Haller embody a medical pattern that holds, with variations, for most physiologists of German-speaking lands in the eighteenth century, including such lesser lights as Johann Theodor Eller, Daniel Bernoulli (*qua* physiologist), Johannes Nathanael Lieberkühn, Georgius Prochaska, and Cristoph Girtanner.[21]

By the end of the century a number of men had appeared in the German-language areas who, while otherwise conforming to the medical pattern, adopted approaches to physiology that were more philosophical and broadly "biological" than those of their predecessors. Among them were Johann Friedrich Blumenbach, Johann Christian Reil, and Carl Friedrich Kielmeyer. Kielmeyer, whom Coleman has called "perhaps the preeminent teacher of physiology in Germany in the generation before Johannes Müller," took an M.D. in 1786 but never practiced medicine.[22] Instead he taught zoology and chemistry at the Karlschule near Stuttgart, and after 1796 chemistry, botany, materia medica, and pharmacy at Tübingen. Though never reaching the extremes of *Naturphilosophie*, Kielmeyer sought to elaborate a global philosophy of nature. Physiology, in his view, was to be based on the comparative study of plants, animals, and man and of the forces underlying their relationships. Its empirical dimension was to be comparative anatomy, not experiment. In the work of such men, German physiology by the end of the century was, if anything, moving away from experiment as a central method of investigation.

In Italy the medical men, who tended to do physiology as an aspect of anatomy or in connection with clinical medicine, included Leopoldo Caldani, Domenico Cotugno, and Luigi Galvani.[23]

More significant for Italian physiology was the work of two nonmedical men, Lazzaro Spallanzani and Felice Fontana. Both were what Claude Dolman has called "priest-polymaths," making their careers through a mixture of church and academic posts, and private patronage. Both were physicists, in the eighteenth-century sense of those who studied nature in an experimental rather than a mathematical way. Both embraced a range of research topics much wider than that characteristic of their medical contemporaries. In particular, both adopted a more broadly biological approach to the study of living things.[24]

Despite their differences, medical and nonmedical groups overlapped in significant ways. Members of both were strongly influenced by Haller's work. Of the five Italian physiologists mentioned above, four were in some way associated with the University or the Istituto delle Scienze of Bologna. None of the members of either group were physiologists in any specialized sense, and only one (Caldani) used the term "physiology" in the titles of publications. Above all, though most members of both groups made some use of experiment in their physiological work, none established an institutionalized research tradition in experimental physiology.

In eighteenth-century Britain, physiology was overwhelmingly the work of medical or surgical men. Some of these, like Alexander Stuart and James Jurin, supported their research with successful medical practices. Others, including William Hewson, John Hunter, and William Cruikshank, found additional training, encouragement, and material support for their scientific work in William Hunter's Great Windmill Street School of anatomy in London. In Scotland the chairs of the university medical faculties offered some opportunity for teaching and research in physiology. Robert Whytt taught the institutes of medicine and the practice of medicine at the University of Edinburgh, and William Cullen held chairs in the same subjects at the University of Glasgow. The major exception to the medical rule, Stephen Hales, followed a Cambridge M.A. in theology with lifelong service as a village curate. Like his Italian counterparts Fontana and Spallanzani, Hales received material and moral support from his church position. Unlike them, he never held an academic post. No one of these men was a specialized physiologist, and the term "physiology" was not in general use.[25]

Though physicians and surgeons played a predominant role in British physiology, important models and motivation came from outside the corporate and learned world of medicine. Early in the century the experimentalism and mechanism of Newton's *Opticks* served as exemplars for some research, including that of Hales, Jurin, and Stuart. Most British physiologists, including Jurin, Stuart,

Whytt, Cullen, Hewson, Hunter, and Cruikshank, were associated with the Royal Society of London. Whatever its limitations in other respects, that institution encouraged experimental inquiry and held up an ideal of scientific knowledge as something of value above and apart from its medical utility. The churchman Hales found inspiration in natural theology, to similar effect. The Scottish universities, where Whytt and Cullen taught and where Hewson and Cruikshank studied, were more open to the modern sciences than their English counterparts. This circumstance helps to account for the disproportionate representation of Scotsmen among British physiologists, a numerical leadership that continued into the nineteenth century.[26]

In Britain, as elsewhere in Europe, experimental control of life processes did not occupy a necessary or central place in physiological inquiry. Just as physiology was most often one among several subjects cultivated by a given investigator, so animal experiment was usually one among several means of investigation employed. For some physiologists, and in some pieces of research, experiment was displaced entirely. Most prominent among alternative approaches were systematic, a priori theorizing, usually in a mechanist mode; inference of function from structure (dead anatomy); and clinical observation.

Even when experiment was employed, its role as means of inquiry must be differentiated from its use as the demonstration or illustration of theories arrived at by other, a priori means. A case in point is the work of the physician Alexander Stuart on muscle function.[27] In this case, as in many others in the eighteenth century, experiment was the humble, if not abject, servant of prior theoretical commitments.

Some British investigators, such as Hales and Whytt, carried out experimental investigations worthy of comparison with those of the next century. But Hales, isolated in his village parsonage, was hardly in a position to turn out a new generation of experimentalists. Whytt, though professor of medicine at Edinburgh, did not. More characteristic of the latter half of the century was William Cullen, whose course on the institutes of medicine at Glasgow emphasized nosology, his profession's answer to the prestige of Linnaean science.

In many respects, physiology in France before 1790 conformed to patterns observed elsewhere in Europe. Physiologists were physicians or surgeons. The most notable exception, Réaumur, was like his nonmedical counterparts Trembley, Humboldt, Spallanzani, Fontana, and Hales in cultivating scientific interests that carried him beyond the medical preoccupation with human structure and function.[28] Livings came from professional practice or teaching. In only one case—the chair of physiology at the Paris Collège de Chirurgie—was a teaching post specifically designated for the subject. That chair and its holders, Antoine Louis and Toussaint Bordenave, were also exceptional in using the term "physiology."[29]

Paris and Montpellier, in contrasting ways, dominated research. Many of the Paris physiologists, including François Pourfour du Petit, Anne Charles Lorry, Nicholas Saucerotte, Jean-Louis Petit, Jacques Tenon, Louis, and Bordenave, were surgeons or connected in some way with surgery.[30] Several, including Pourfour du Petit, Réaumur, J.-L. Petit, Tenon, and Saucerotte, were members or correspondents of the Academy of Science. Lorry was a leading member of the scientifically progressive Société Royale de Médecine. The association of surgeons and science gave experimental methods a relatively favored place in the physiology of the capital. All of the individuals mentioned above did experimental work, in some cases with notable results.

Antoine Louis, to take only one example, performed animal experiments on the physiological mechanisms of violent death. In his studies of drowning, he aimed to discover unequivocal signs of death and to find means of reviving drowned persons. When death occurred by strangulation, he wished to know whether suicide or homicide was involved. In both cases the researcher's goal was practical, but the means included experiments on living animals and human cadavers, and physiological knowledge was a by-product. Bichat would later turn study of the mechanisms of violent death more directly to physiological ends. Louis also initiated investigations of the analogies of structure and function of the pleura and pericardium and of their pathological alterations, of the kind later to be carried much further by Bichat in his tissue studies.[31]

With its ancient university and Mediterranean locale, Montpellier had long been an important center for medicine and science. The eighteenth century marked a high point of its activity and reputation. The sciences were vigorously pursued in the Medical University, in the Collège de Chirurgie, and in a new

Société Royale des Sciences, with physicians playing a prominent role.[32]

In the physiology produced by Montpellier savants, discursive theorizing predominated. Where empirical work was included it tended to be human or comparative anatomy, with the place of experiment at best subordinate or peripheral. What distinguished Montpellier physiology and gave its authors collectively the attributes of a school were its well-argued critiques of mechanism. These were joined to concepts of the specifically vital in living things that made use of analogs to Newtonian gravitation. François Boissier de Sauvages, a physician-demonstrator at the Jardin des Plantes, thought the key to medicine lay in nosology, or the classification of diseases according to their symptoms. He went on to argue the need for a "philosophical nosology" that would explain the bodily processes underlying disease states by faculties peculiar to the living body. These faculties were no less efficacious for being, like gravity or elasticity, unknown in their essence.[33] The professor of medicine Théophile Bordeu advanced a concept of sensibility as something distinct both from mechanical forces and from conscious mind or soul. Immanent in all living fibers, it was also differentiated and decentralized, taking a form in each organ appropriate to the control of its function.[34]

These trends culminated in the writings of Paul-Joseph Barthez, who held the chair of anatomy and botany at the Medical University from 1785 to 1792. Barthez sought a foundation for medicine in a physiology—or, as he termed it, a "new science of man"—that would occupy itself with phenomena found exclusively in the living body. His methodological stance was explicitly Newtonian: we cannot know the essence of causes, but only the regular lawlike relations among phenomena; phenomena governed by the same laws can be considered effects of the same cause (cf. Newton's Rule II, *Principia*); and if laws governing similar phenomena are different, those phenomena must be considered the effects of distinct causes. Like Stahl and Sauvages, Barthez saw a need to posit a faculty that would coordinate and direct bodily actions. This was the *principe vital*, unitary and ontologically distinct from both mind and mechanical forces, known through its effects but unknown in essence.[35]

Though his physiology did not gain a preeminent position in the French medical world of the later eighteenth century, Barthez's ideas did have consequences. It was his use of *principe vital* that brought the term "vitalism" into use toward the end of the century.[36] Barthez's ideas figure in Pierre Cabanis's reflections on certainty in medicine on the eve of the Revolution.[37] Barthez's influence is evident in Bichat's program for construction of a scientific physiology and in aspects of his vital properties doctrine. It may be seen in Magendie's early programmatic position. There is an unmistakable resonance between the methodological writings of Barthez and those of Claude Bernard, and it may be that Bernard drew upon Barthez's Newtonian conception of physiology.[38] Barthez's writings reflected and reinforced the integration of physiological and medical thought in France.

Prior to 1790 French physiology displayed much the same heterogeneity and lack of differentiated institutional status visible elsewhere. What distinguished the French case and prepared the changes that began in the 1790s was an institutional context—already in place by the beginning of the Revolution—that would provide the new field with vital knowledge, techniques, locales and exemplars for research, and material support.

The leading center of scientific professionalism in Europe, the Paris Academy of Science, though it allotted no place to physiology as such, had from the beginning of its history included physiological investigations under the rubric of anatomy or other subjects.[39] A similar disguised and partial institutionalization of physiology had occurred at the institutions founded and sustained by royal support since the sixteenth century. At the beginning of the eighteenth century the chairs at the Collège Royal included those of medicine, surgery, pharmacy, and botany, each of which at some time undertook investigations that would later be termed physiological. The Jardin du Roi possessed a chair of anatomy from the mid-seventeenth century. Its first demonstrator, Pierre Dionis, taught human anatomy "following the circulation of the blood and the recent discoveries."[40] More important than the particular investigations carried out under their auspices was the commitment to basic research that these institutions embodied, and that provided a counterweight to the practical preoccupations of physicians and surgeons.

Equally portentous was the progress toward professional status accomplished separately by French surgery, pharmacy, and veterinary medicine by 1790. Medicine had attained the status of a liberal profession in Paris in the thirteenth century. Indeed, medicine can be considered the archetype of the liberal profession in the modern West, since from this time it was in possession of the elements usually associated with professionalism: an esoteric body of knowledge; institutions for the transmission of that knowledge; exclusive control over admission of new members to the corporative body; and official recognition by the state. Originating with the universities, the medical profession took over the commitment of medieval Latin culture to a fundamental distinction between the work of the mind and the work of the hand. The former, considered to be of higher value, was the province of the learned professions and clergy. The latter was relegated to the socially inferior "mechanical" arts. In the medical realm this distinction took the form of a radical separation of medicine (conceived as theoretical knowledge of the human body and practical knowledge of regimen and drugs) from surgery (conceived as the practice of manual operations—especially phlebotomy and bonesetting—carried out directly on the body). Medical doctors received university training and formed a small, elite, and powerful corporation. Surgeons or barber-surgeons were trained by apprenticeship, and despite early efforts to emulate the medical doctors were for a very long time unsuccessful in developing their own effective organization.[41]

Surgery was transformed between 1650 and 1750 in a gradual process that led, against the resistance of the medical faculty, to the surgeons' achievement of the status of a liberal profession.[42] While imitating the organizational and honorific patterns of physicians, the surgeons retained an emphasis on the "union of hand and mind," a fact of decisive importance for the transformation of medical education during the Revolution. As part of the theoretical training offered students, the Paris Collège de Chirurgie, founded in 1724, progressively expanded its courses of instruction in the auxiliary sciences.[43] As already noted for the case of physiology, the effect was to expose the younger generations of surgeons to the most recent developments in fields of basic science and, at least in some cases, to provide an institutional locale for research in those fields.

Pharmacy underwent a very similar evolution. The eighteenth century proved to be a period of transition from traditional to modern patterns of education and professional structure. Private courses of instruction, which had begun to supplement apprenticeship training in the seventeenth century, underwent a rapid expansion from around 1700. Public courses in chemistry and—over the objections of the Paris Faculty of Medicine—botany were introduced. In 1780 the king issued statutes determining the requirements of instruction at the new Collège de Pharmacie, founded in 1777. These included the permanent establishment of public courses in chemistry, botany, and natural history. By the time of the Revolution, an emerging profession of pharmacy had linked its aspirations closely to the modern sciences, particularly analytical chemistry.[44]

The first veterinary schools in Europe were founded at Lyons in 1762 and at Alfort near Paris in 1766. In spite of humble origins, resistance from the powerful corporation of blacksmiths, and the suspicions of physicians and surgeons, veterinary medicine by the time of the Revolution had established the patterns of professional education and activity that still characterize it. Even with its practical emphasis, the program of instruction at Alfort included anatomy, physiology, hygiene, zootechnics, and artistic anatomy. In the 1780s instruction at Alfort was transformed by the nomination of prestigious professors, the creation of new chairs (1784), and the initiation of advanced research in the sciences. Antoine Fourcroy taught chemistry, Louis Daubenton rural economy, Pierre Broussonnet zoology, and Félix Vicq d'Azyr comparative anatomy. Veterinarians were increasingly called on by the government for consultation and to carry out studies of special problems. The primary motive of the crown in supporting the schools was its hope to control epizootic disease in the countryside, and insufficient funds soon forced a reduction of the scientific content of instruction. Nevertheless the ideal of a unified human and animal medicine held by Vicq d'Azyr and the veterinarians was taken over by the Société Royale de Médecine and made part of its plan of medical reform. Scientific research continued to be carried out at Alfort through the Revolution and beyond.[45]

In their strong and successful movement toward professionalization in eighteenth-century France, the healing arts derived advantages of intellectual prestige and increased effectiveness from their association with the sciences. In return the sciences received

benefits that were no less important. Though increasingly specialized and esoteric as fields of knowledge, the "auxiliary" sciences
found new support and protection by being embedded in institutions providing direct practical services to society. Teaching positions in the new institutions of professional education provided
a means for diffusing the latest scientific knowledge to new generations of students, some of whom might now be recruited to
the sciences. Though the opportunities were not always utilized,
such positions could also enable their holders to do original research. For physiology the benefits were partly direct and immediate, as in the case of the work of Louis and Bordenave at the
Collège de Chirurgie. To a far greater extent they were indirect
and delayed, and derived from the impetus given to the development of new knowledge and techniques in closely related sciences such as chemistry, anatomy, botany, and pharmacology. Not
until the first two decades of the nineteenth century would the
significance of these developments for physiology become fully
evident.

Research and teaching of the sciences associated with medicine
was also developing outside the existing institutional structure, in
the private or "free" courses organized by individual savants.[46]
Though precedents go back as far as the sixteenth century, it was
only in the second half of the eighteenth century that this form
of instruction took on its full development. The free courses helped
fill the large gaps left by overly traditional official instruction, and
allowed the development and diffusion of knowledge in new or
expanding subject areas, such as anatomy, physiology, chemistry,
surgery, and obstetrics. The abandonment of the medieval *lectio*,
the replacement of Latin by French, and the smaller, more intimate
setting made for better rapport between teacher and students. The
private courses were generally more practical in structure and
content. They made more frequent use of cadavers and of demonstrations and experiments, and used the clinical approach in
medical subjects. In general, the professor had at least a professional training, if not an institutional position, in one of the healing
arts. Expanding rapidly in the later eighteenth century, the private
courses came to play a major role in the teaching and advancement
of medicine and the accessory sciences on the eve of the Revo-

lution. They would maintain this role through the 1790s and beyond, not least as the locale of the physiological work of Xavier Bichat and François Magendie.

The eighteenth century had prepared the way for a transformation of the content and status of physiology. The transformation itself would come only in the special intellectual and institutional environment created by the Revolution and its aftermath.

We must put aside this excessively hypothetical physiology in order to obtain one more worthy of the men who will be formed in the new schools that we propose to establish.

—*Jacques Tenon* (1794)

2 Context for Change: The 1790s

Reflecting on the place of physiology in the new medical education, the surgeon Jacques Tenon gave concise expression to the conjunction of institutional and attitudinal circumstances that were to make the 1790s a pivotal decade for that science. The commitment to basic research embodied in institutions such as the Academy of Science, the Jardin du Roi, and the Collège Royal was continued, in some cases after a temporary lapse. Eighteenth-century trends toward the professionalization of the healing arts, and their association with the sciences, were consolidated. More obviously connected to the political events of the decade was the establishment of a new medical education that ended the centuries-old separation of medicine and surgery, and centered training and practice in the hospitals. With the new medical education came parallel efforts at the reconstruction of medical theory. Scientifically ambitious physicians beginning their careers in this milieu sought to replace an older medicine, perceived as excessively hypothetical in theory and ineffective in practice, with one drawing on the certainty and utility of the modern sciences. From them would come the first experimental physiology of the next century.

In December 1788, on the eve of the Revolution, the young Parisian physician-philosopher Pierre Cabanis completed an essay in which he proposed to examine "whether, by observation and by simple reasonings which can be immediately deduced from it, a solid basis can be given to the principles of medicine; or whether it is true that the reproaches of uncertainty that several philosophers have made against this art are really justified.[1] Both in his statement of intent and in the title he chose for his essay, *Du degré de certitude de la médecine*, Cabanis expressed a concern with the very foundations of medical theory and practice, a concern that was to become fully self-conscious in the next decade throughout the medical world of Europe. Nor were the issues of academic interest only. Cabanis wrote at a time when the worth and effectiveness of medical knowledge and technique, and therefore of the medical profession itself, were being called into question. The criticism came not only from philosophers in pursuit of some ideal epistemological rigor, but from the public. It came first of all from that very Paris intelligentsia of which Cabanis was a member, and even from some members of the medical profession. To a perceptive observer, medicine might indeed appear to be on the verge of a crisis, in the course of which—following the medical metaphor—it would either recover and improve or, as some extremists predicted, die. Cabanis experienced this situation in all its gravity, and responded with a diagnosis of the illness and a prescription for therapy. His essay of 1788 provides a window on the inner condition of European medicine on the eve of the French Revolution.[2]

With the outbreak of the Revolution the following year, the crisis of medicine was abruptly forced beyond professional self-doubt and fashionable criticism into the arena of political and institutional change. The Paris Faculty of Medicine fell, never to be revived in its old form. So too did other institutions associated with the healing arts, though their demise was to be only temporary, and much of their substance, if not their names, was to survive the radical phases of the Revolution. War and quackery forced the first steps of reconstruction. What had been ideas in the minds of reformers in the Société Royale de Médecine and elsewhere became blueprints for a new system of medical education, and from materials prepared by the eighteenth century a new Ecole de Santé took form. After Thermidor the dominant mood in medical circles was one of renewal and fresh beginnings. A new generation of doctors appeared—enthusiastic, optimistic, ambitious, talented. Paris was launched on a half century of world medical leadership.

None of this was visible to Cabanis as he wrote in 1788, though he was to have a prominent role in the events of the next decade. What he saw and responded to in *Du degré de certitude* were criticisms of the content of medical theory and practice as it was in the 1780s. Those must have been difficult years in which to begin a medical career in Paris. The Faculty of Medicine was visibly in decline, and the medical profession the object of widespread mistrust. Though attacks on medicine and the medical profession were no novelty in educated French circles, long familiar with the skeptical essays of Montaigne and the caricatures of Molière, they appear to have taken on new intensity and conviction in the last decades of the eighteenth century. One source of antimedical sentiment was Rousseau, whose novel *Emile* portrayed medicine as yet another artificial obstacle placed by corrupting civilization in the path of a benign nature. Even medicine's partisans had to admit that the uncertain and confused condition of medical theory made a striking contrast with the magisterial progress of the mathematical sciences since Newton, and with the brilliant advances in chemistry which followed one another in quick succession from the 1770s. Cabanis, living at the center of Paris intellectual life, was acutely aware of both tendencies.[3]

In his response to the criticism of medicine Cabanis juxtaposed cautious affirmation of past achievements with prescriptions for reform. To counter the charge that physicians were ignorant of the causes of life, disease, and the actions of medicines, he invoked the distinction between knowledge of cause or essence and knowledge of the regular relations of phenomena accessible to the senses. Here his language was reminiscent of, and perhaps directly borrowed from, Barthez. The neo-Hippocratic empiricism that informed many of Cabanis's statements receded in favor of a view of scientific and medical knowledge derived ultimately from Newton via Locke and Etienne Condillac.[4]

Cabanis appealed especially to Condillac to show that even the complex phenomena of disease—like the complex constructions of language—may be analyzed into a relatively small number of simpler elements. Here Cabanis asserted as an article of faith and a desideratum what Philippe Pinel would attempt to put into practice nine years later in his *Nosographie philosophique*.[5]

Cabanis did not offer a critical discussion of the techniques and pitfalls of animal experiment, but only some general considerations on the degree of certainty attainable by medical knowledge. This reflected not the paucity of experimentation in past and current medicine—examples were available—but the bookish and literary character of his own approach to the problems. His answer to critics who pointed to disagreements among physicians was likewise programmatic rather than concrete, calling for a self-conscious empiricism and more modest theoretical ambitions.[6]

Whatever the imperfections of the healing art, Cabanis concluded, people would always seek relief from suffering. If medicine did not exist, it would have to be invented. If the physician often did little good, or even some harm on his own account, by displacing gross incompetents he prevented much evil. That was his true role.

Cabanis's essay was completed in 1788 but not published until 1798. It therefore reflects one observer's perception of the situation of medicine in France prior to any of the political events of the 1790s, without itself contributing directly to the climate of opinion of the early phase of the Revolution. Some of the same themes were widely sounded in French medical and scientific circles in the early 1790s. Pierre Desault, the great surgeon of the Hôtel-Dieu and teacher of Bichat, believed internal medicine to be obscure, speculative, and unscientific, and opposed the unification of medical and surgical instruction at the Ecole de Santé.[7] Similar reactions are found among the medical reformers. Physiology, as the basis of medical theory, was often singled out for special criticism. The Société Royale de Médecine's *Nouveau plan de constitution pour la médecine en France*, drawn up in 1790 by the society's president, Vicq d'Azyr, called for a physiology closely tied to gross anatomy and observable function: for "anatomy may be separated from physiology; but physiology cannot exist alone; it must be joined to the study of the human body, without which it will always wander from system to system."[8] In the Comité de Santé's debates on the question of whether the new curriculum should include a combined course of anatomy and physiology or whether each course should be taught separately, the very existence of physiology was questioned. Antoine Louis noted that some fellow delegates thought of physiology as merely "a romanticized form of anatomy."[9]

The criticism of internal medicine in general, and of physiology in particular, voiced by physicians and surgeons in the 1790s reflected in part prevailing attitudes in French scientific circles. These attitudes can be found in a report to the National Convention in 1795 by Antoine Fourcroy, in which the prominent chemist referred to the old medical education as involving the teaching of "useless physiologies."[10] Jacques Tenon, a delegate from the Academy of Science, expressed similar views in testimony given to the Comité de Santé. Tenon was a distinguished surgeon and anatomist, a skilled diagnostician, an authority on hospitals, and a leading figure in the medical reform movement. In his testimony to the committee, Tenon drew a sharp contrast between anatomy, an exact science based on strict visual inspection, and physiology, also a science but one that relied mostly on hypotheses. He maintained that it would be a mistake to combine the teaching of the two subjects, since "that would in most cases result in mixing certitude with doubt, truth with error." Anatomy was "positive" and an essential basis of practice, especially in surgery. Physiology was diffuse and hypothetical, and difficult to relate to practice.[11]

Contrasting the evident progress of anatomy since the Renaissance with the relative underdevelopment of physiology, Tenon called for the construction of a new physiology based on animal chemistry (still largely unknown), mechanics, applied mathematics, and animal experiment. Above all, professors of physiology in the new medical schools should be able to relate their science to pathology and to make it inform medical practice. Here Tenon might have pointed to his own investigations in the 1750s on bone exfoliation. Occasioned by clinical observations, those studies had relied solely on rigorous gross anatomical analysis and animal experiment, excluding speculation on underlying mechanisms. The result had been an addition to physiological knowledge, but one in which science had been subordinated to the illumination of a clinical and pathological phenomenon.[12]

As Tenon's own work shows, such oversimplified views do injustice to the real achievements of the physiology of the preceding two centuries. That they were held and expressed with great conviction by leading medical and scientific figures is, however, highly significant for the definition of the contemporary climate of opinion in France. The debates and programs of the early phase of the Revolution revealed that to the best minds medicine—including its fundamental theoretical component, physiology—was vulnerable to charges of uncertainty at its very foundations.[13]

Before Cabanis's essay could be published, medicine—along with all other facets of French life—was swept up by the events of the Revolution. So far as medical, scientific, and educational institutions are concerned, the Revolution proceeded in two broad phases, divided roughly by Thermidor. The first phase, running from the convening of the Estates-General to the fall of Robespierre, witnessed extensive dismantling of the institutions of the Old Regime under the pressure of radical liberal and democratic ideology. The Faculty of Medicine was dissolved in August 1792. For the sciences, the ideological rejection of aristocratic forms of corporate organization culminated in August 1793 with the abolition of the learned academies of France, including the Academy of Science, as incompatible with a republic. The return to political order and centralization of power began in December 1793 with the establishment of the revolutionary dictatorship and the Terror. By early 1794 the practical pressures of a wartime situation were beginning to outweigh ideological suspicion of scientific and technical elites. It was not until after Thermidor, however, that the reconstruction of French medical and scientific institutions by the Convention began in earnest. The Ecole Centrale des Travaux Publiques (later the Ecole Polytechnique) was founded in September 1793, the Conservatoire des Arts et Métiers and the Ecoles Normales in October, and the Ecoles de Santé in December. February 1795 saw the creation of the Ecoles Centrales, and in June the Bureau des Longitudes came into existence. The Academy of Science was revived as the First Class of the Institute in October. The attendance of all five members of the Directory, along with other government dignitaries, at the first public meeting of the Institute signaled the reestablishment of close ties between learning and government, the intellectual and governing elites.[14]

With the outbreak of the Revolution the last remaining strength of the Paris Faculty of Medicine, its status as a long-established and venerable institution, became a weakness. In the altered political situation, polemical attacks and failure of public confidence were translated into concrete legislative measures against the medical establishment. In August 1792 all university faculties and medical schools in France were abolished by law. Medicine, some legislators thought, could now return to a primitive, natural state, and eventually to the conditions of ancient Greece. Legal prerequisites for medical practice were abolished and the medical profession was fiercely attacked. One speaker in the Convention argued that "doctors should be dealt with like priests—both alike being mere jugglers." In reality, the door had been opened to quackery and charlatanism. As Cabanis had predicted in his yet unpublished essay, a greater number of victims were now delivered into the hands of "audacious ignorance."

In the suffering brought about by the anarchy of medical practice the public began to discover for itself, if not the positive value of the older medicine, then its negative utility as Cabanis had described it. Beginning in 1792, public opinion and pressure gradually built up in favor of the reorganization of medical teaching. This pressure was powerfully reinforced by the needs of the army and navy, which lost over six hundred doctors in the wars between September 1793 and the end of 1794. In the meantime surgical institutions, in spite of their superior reputation under the Old Regime, had also succumbed to revolutionary pressures. The Paris Académie de Chirurgie persisted legally, but declined in activity, until its abolition in 1794. Projects and blueprints for the reform of medical education proliferated. With the relative relaxation of political pressures after Thermidor, covert introduction of semi-private instruction in the hospitals became widespread, indicating discontent with the absence of professional standards. By the summer of 1794 there was general recognition that although the old medicine was gone for good, nothing satisfactory had taken its place. The time had come for the reconstruction of medical education.[15]

A few weeks in advance of Thermidor—in July 1794—the Convention began to consider the problem. The Committee of Public Safety charged Antoine Fourcroy and François Chaussier with the preparation of a plan for an *école revolutionnaire* of the art of healing. Fourcroy, as chemist, physician and member of the Convention, had long been involved in problems of educational reform, and his initiative had set the legislative process in motion. Chaussier had been a prominent figure in medical circles in Dijon, where he had taught chemistry, anatomy, and legal medicine. The events of Thermidor delayed the response of the Convention to the proposals submitted by Fourcroy and Chaussier until the end of 1794.[16]

The final decree embodying these proposals, passed by the Convention on 14 Frimaire an III (December 1794), created the foundations of modern medical education in France. Article 1 declared that "there shall be established a School of Health at Paris, at Montpellier and at Strasbourg; these three schools shall be destined to produce officers of health for the service of the hospitals." This represented a high degree of centralization in comparison to the twenty-odd medical faculties of the Old Regime, although Chaussier, favoring the Jacobin principle of concentration of power, had at first argued for creation of a single school in Paris. The utilitarian goal of the new schools was a response not only to the pressing practical needs of the moment but also to revolutionary ideology which glorified useful knowledge while remaining suspicious of purely theoretical pursuits. Article 4 specified that, in addition to study through reading and lectures, "students shall practice anatomical, surgical and chemical operations; they shall observe the nature of diseases at the bedside of patients, and shall follow their treatment in the hospitals near the schools." This contrasts sharply with the pre-Revolutionary style of teaching at the Paris Faculty of Medicine, which had relied heavily on the medieval *lectio*. It was, however, entirely consistent with the patterns of teaching developed over the preceding decades at the Collège de Chirurgie, and indicates the dominance of the surgical element in the unification of medical and surgical instruction favored by Fourcroy and Chaussier. Article 5 provided for specific numbers of permanent professorships and adjoints at the three schools. This measure would ensure continuity in teaching, another feature of the Collège de Chirurgie the lack of which had been a major weakness of the Paris Faculty of Medicine. Finally, Article 13 established an annual stipend of 1,200 livres over a period of three years—the length of a complete course of studies—for each of the 300 students to be admitted. Before the Revolution, medical training had been so expensive that all those lacking wealth or connections had found it extremely difficult to obtain— a circumstance from which Fourcroy himself had suffered—and provision of a stipend gave expression to the Revolutionaries' commitment to make specialized training available to a wider social group.[17]

In the organization of the Ecole de Santé of Paris, a single chair was allotted to anatomy and physiology. On 31 January 1795 Chaussier himself was named to this chair, which he was to hold until the 1822 purge of the Faculty of Medicine. Chaussier's first adjoint, named at the same time, was Antoine Dubois. A former student and assistant of Desault and Jean Baudelocque, Dubois had been the last professor of anatomy at the Collège de Chirurgie's Ecole Pratique to be named by Louis XVI, in 1780.

Begun in the 1750s, the Ecole Pratique had annually selected twenty-four of the best students from the public courses of the Collège for advanced experience in anatomical dissection and surgical operations on cadavers. This training was highly valued by aspiring surgeons. By the time of the Revolution the Ecole had four professors teaching twenty-six students, and the annual period of instruction had been extended from four to six months.

Dubois, who left the chair of anatomy and physiology in June 1795 to join the clinic of the Hôpital de Perfectionnement, played a major role in the reopening of the Ecole Pratique in 1797. Through the new Ecole Pratique, open to 120 select students by competition, the Paris Ecole de Santé became a center of research as well as instruction. Under the direction of Dubois, Dupuytren, and others, researchers carried out studies combining anatomical, anatomical-pathological, chemical, and vivisectional approaches.[18]

The Collège de Pharmacie, though subject to many of the same stresses as the Faculty of Medicine and the Collège de Chirurgie, had a smoother passage through the Revolution. The National Assembly's decree of 2 March 1791 suppressing corporations and proclaiming the liberty of professions and trades included pharmacy in its scope. The dangers inherent in an unregulated traffic in drugs were quickly apparent, however, and the provisions of the decree bearing on pharmacy lasted barely six weeks. By a special decree of 17 April 1791 the National Assembly ruled that the statutes and regulations in force before March 2 were to be reinstated and that the practice of pharmacy was to be restricted to those meeting the requirements of the profession.[19]

Nevertheless a corporation dating from the Old Regime could not feel entirely secure in the radical democratic and anticorporative atmosphere of the 1790s. On 20 March 1796 the members of the Collège de Pharmacie, taking advantage of the law of 23 September 1795 giving citizens the right "to form private establishments" for education and teaching, and free societies "to co-operate in the progress of the sciences, letters and the arts," reconstituted themselves as the Société Libre des Pharmaciens de Paris. The first courses of the new-old Ecole Gratuite de Pharmacie opened in 1797.[20]

The final metamorphosis of the old Collège de Pharmacie, and the one that gave it its distinctively modern form, was embodied in the legislation of 21 Germinal an XI (11 April 1803). Here the major changes had to do rather with organization than with the content of teaching or practice. State control was substituted for the old corporate independence in matters of training, certification, and policing of the profession. A single system of pharmacy for all France was instituted, and three écoles de pharmacie were established as training centers in Paris, Montpellier, and Strasbourg to parallel the schools of medicine.

The centering of medicine and surgery in the hospital in the course of their reorganization in the 1790s had its parallel in the creation of a hospital pharmacy. The Apothicairerie Générale, installed on 18 Prairial an III (6 June 1795) was replaced by the Pharmacie Centrale des Hospices on 22 Brumaire an V. Two decades later, in 1815, the internship in pharmacy was introduced. These developments had been largely foreseen and supported in the Société Royale de Médecine's *Nouveau plan* of 1790. This had called for the placing of a pharmacy in the buildings of the proposed new school of medicine, which itself was to be annexed to a hospital, so that the preparation, administration, and distribution of medicines would be integrated with the teaching of medicine and with the medical practice of the hospital.[21]

The orientation of the pharmacists of Paris toward the utilitarian and democratic values of the day and their sense of civic responsibility are reflected in their relationships with the local authorities of Paris. It was at the initiative of a Paris pharmacist, Charles de Gassicourt, that the first Conseil de Salubrité (Health Council) was formed, a panel of experts to be at the regular disposal of the prefecture of police. The original Conseil later developed into the Conseil d'Hygiène Publique et de Salubrité of the *département* of the Seine, which became a model for similar committees in all the French departments.[22]

The veterinary schools of Alfort and Lyons traversed the period of the Revolution without major change. At the time of the *Nou-*

veau plan it appeared that veterinary medicine might be integrated in some larger, more inclusive institutional framework. The president of the Société Royale, Vicq d'Azyr, favored the grouping in a single establishment of the faculties of medicine, the schools of surgery, the Jardin du Roi, the Jardin des Apothicaires, and the veterinary school of Alfort. Tallyrand, in his report of 1791 to the Constituent Assembly, supported the conception of a unitary human and animal medicine, a program advocated in England by John Hunter. It was not the ideas of scientists or reformers, however, but the outbreak of war that determined the position of the veterinary schools. While remaining organizationally distinct, their importance and prestige were heightened by the suddenly increased demand for horses and for technicians qualified to maintain the cavalry. Scientific activity was not abandoned in the face of practical demands. The momentum of the academic period of the 1780s carried over into the 1790s and beyond. The first holder of the position of Chef des Travaux Anatomiques at the new Ecole de Santé de Paris was Alfort's Honoré Fragonard. Jean Girard, a graduate and subsequently professor and director of Alfort, collaborated with Chaussier in the latter's formulation of a new anatomical nomenclature based on insertions. Before his early death in 1796 Pierre Flandrin, a professor at Alfort, carried out and published important experimental studies of lymphatic circulation and intestinal absorption.[23]

With the exception of the old Faculty of Medicine, the major institutions of the healing arts in France weathered the Revolutionary storm with a high degree of resiliency and continuity. Even the écoles de santé were heavily indebted in course content, faculty, and building facilities to the old Académie and Collège de Chirurgie. In the period of national reconstruction of the Directory, Consulate, and Empire, the major permanent institutional innovations derived from measures of rationalization, unification, and centralization under state control that represent as much the culmination of Old Regime tendencies as the creation of the institutions of modern France.

One institution in which significant internal change did occur as a result of the Revolution was the Academy of Science. The change was less in organization or principles, than in the function the Academy, now the First Class of the Institute, played in the scientific community. In the period of over two years from the dissolution of the Academy to the opening of the Institute, scientists found refuge and new support in the more specialized scientific institutions that had been renewed or created in the interval. In the process they tended to find more narrowly defined identities: the former generalist member of the Academy became a physicist at the Ecole Polytechnique or a zoologist at the Muséum National d'Histoire Naturelle. One result was that the First Class tended to become the agency through which original research in established disciplines was recognized and recorded, after this research had first been created and judged elsewhere. Not included in the trend were subjects not yet clearly constituted as sciences and lacking separate institutional locales. In these cases the First Class could still provide an invaluable forum for debate and creative activity. Meteorology was one such subject in need of institutional support. Physiology would soon be another.[24]

It was Fourcroy's plan for a national institute that was finally adopted and put into practice in 1795. Wishing to avoid a permanent state of conflict within the Institute between the scientific areas on the one hand and the technical or applied areas on the other, Fourcroy introduced into the First Class a number of sections devoted to medicine, surgery, rural economy, and the veterinary arts. This action reflected first of all Fourcroy's commitment to a close association between the sciences and the healing arts, expressed several years earlier in the first number of his journal, *La médecine éclairée par les sciences physiques*.[25] It also gave expression to the long-developing scientific aspirations and achievements among the practitioners of the healing arts. At the same time the scientific movement within the healing arts was given new impetus and direction by the provision of a definite institutional context in which it could proceed. Younger workers in medical fields had an appropriate institutional target for their scientific ambitions. Those medically trained people who did become members of the First Class entered a milieu in which they constantly mixed with leading scientists and were encouraged to adopt or strengthen generally accepted scientific standards and attitudes. For some of the sciences associated with medicine, institutionally weak and struggling to establish their intellectual credentials, the new foothold within the First Class would prove decisive. One such science was physiology.

Thermidor was followed by a movement of reconstruction proceeding simultaneously in all areas of French life—political, social, military, and intellectual. The mood was expansive: with the foundations cleared by cataclysm, the way now seemed open for the reassertion of Enlightenment optimism and faith in reason in the building of a new and better edifice. Intellectuals, as the principal bearers of this faith, and who themselves had been temporarily eclipsed by the Revolution, joined the movement of renewal with enthusiasm. Nor was this merely a matter of self-interest: many believed that a return to philosophy, the cultivation of the mind, would itself be a force for social healing and renewal. At the same time, constructive change was fostered by a reordered system of social values in which merit became the chief criterion of social advancement: the "career open to talent" replaced the Old Regime privileges owing to birth, and the Jacobin tendency to deny all differences of native capacity in the name of equality. The new social mobility is well illustrated in the field of medicine, where the process was aided by the active and perceptive patronage exercised by the older professors. Its beneficiaries were to include such figures as Dupuytren, François Broussais, Matthieu Orfila, Jean Esquirol, and Bichat, all beginning their careers in this period.[26]

Nowhere was this movement of reconstruction more badly needed than in the field of medical theory. By destroying the old Faculty of Medicine and creating a situation in which widely diverse ideas on medical theory and practice freely competed, the Revolution clearly exposed the internal crisis of medicine. Deprived of its official status and institutional support, the older medicine, already weakened by attacks from without and self-doubt within, proved unequal to the new challenges. When, after the demise of the Faculty, it became evident that the policy of laissez-faire applied to medicine was not the panacea that some had expected, the opportunity—and the need—to rebuild medicine on new foundations presented itself. The écoles de santé embodied the institutional response to this challenge. The theoretical response—partly reflected in the course content of the new écoles—made a clear break with orthodox medical thought in order to draw extensively on the methodological and positive heritage of the eighteenth century.

Cabanis was the most prominent figure in the methodological and programmatic movement for the reconstruction of medical theory in France. After completing *Du degré de certitude de la médecine* in 1788, Cabanis had become actively involved in the politics of medical reform.[27] His consciousness of the crisis in medical theory and practice and his articulation of its elements became available to the medical world of France and Europe when, in February 1798, he presented *Du degré de certitude* to the Institute. Undoubtedly Cabanis felt that his ten-year-old essay still addressed live issues. In addresses to medical students in Paris in 1797 Cabanis emphasized his view that medicine was in the midst of a great revolution, and that its future rested with the generation to which he was speaking. The prominence of the author and the timeliness of his arguments ensured the essay a wide readership in reforming medical circles. The young Société Médicale d'Emulation, which in 1798 published the works of Bichat, Pinel, Anthelme Richerand, and Pierre Roussel, claimed to ally medicine and philosophy following the author of *Du degré de certitude*.[28]

Cabanis's thinking on the situation and prospects of medicine had largely matured by 1788, and his position was not fundamentally altered by the events of the 1790s. With other *idéologues*, he was a product of the Enlightenment. Nevertheless the Revolution did give special prominence and influence to certain Enlightenment themes in both politics and medicine. Denigration, in medicine, of metaphysics and systems in the name of experience paralleled politicians' attacks on tyranny in the name of liberty. Defense of empiricism in medicine by appeal to classical, especially Hippocratic, models provided the medical analogue to the use of Greek and Roman history by the revolutionaries. Both medical and political reformers shared a fervent optimism: perfection was possible, and might be imminent. In Cabanis's thought, these parallels became identities: medicine, conceived as the all-embracing science of man, would show the way to future perfection by revealing the intimate interdependence of the physical and the moral.[29] No doubt this perception of medicine as the key to social renewal intensified the urgency of medical reform in the eyes of Cabanis and some of his contemporaries. It did not, however, significantly change their characterization of the nature of the crisis facing medicine, or of the manner in which this was to be overcome.

In *Du degré de certitude* Cabanis had responded to the contemporary skepticism regarding medicine—both philosophical and popular—with the formulation of a version of medical empiricism drawing heavily on Barthez, Condillac, and the Hippocratic corpus. In essentials, this methodological stance was consonant with the dominant intellectual climate in France after Thermidor. Henri Gouhier, in his intellectual biography of Auguste Comte, has termed the climate of these years "prepositivist," viewing it as the culmination of post-Cartesian trends in French thought that gathered special focus and intensity in the second half of the eighteenth century. Several of the terms later used by Comte to specify the senses of the positive—"precise," "certain," "useful," "real"—also evoke the ideal of knowledge widely accepted in French thought at the end of the century.[30]

One source for the prepositivist intellectual climate of the Directory was the long-established philosophical position of the Paris Academy of Science, recently resurrected as the First Class of the Institute. Faced with disagreements among its members on points of theory, the Academy had from an early date resolved not to commit itself as a body to one or another controverted view, but to assert as definitely true only those points on which general agreement had already been reached.[31] In practice this resulted in a kind of phenomenological positivism, since general agreement was most often accorded only to factual statements summarizing observational or experimental results. Bernard de Fontenelle had been one of the foremost spokesmen for this policy within the Academy. With the new cultural ascendency of the First Class of the Institute and the high social prestige enjoyed by scientists from the Directory on, an opening was created for the wider diffusion of this philosophical position.

Some of the materials available for the task of reconstruction are indicated in the introductory sections of the first issue of Fourcroy's journal, *La Médecine éclairée par les sciences physiques*, which appeared in Paris in 1791.[32] Writing in the atmosphere of high optimism and confidence of the early years of the Revolution, Fourcroy proposed to bring together the results of meetings of a free society of savants interested in the unification of medicine, surgery, and pharmacy. His motives are revealing of the condition and opportunities of medical theory in the 1790s. The enormous growth of the sciences in scope and technicality over the past century had made specialization inevitable, with the result that the old generalist posture of the scientifically oriented physician had now become untenable. No practicing physician at the end of the eighteenth century could hope to have the thorough knowledge of the chemistry, botany, and physics of his time that Boerhaave had had at the beginning of the century. At the same time, partly in response to the increasingly esoteric character of the sciences and partly in reaction against the superficial and overly deductive application of physics and chemistry to medicine carried over from the seventeenth century, many physicians had come to doubt the relevance and usefulness of the sciences for medicine. Fourcroy's aim was to reverse this trend by making available to the busy practitioner bare, concise statements of recent discoveries in the various fields related to medicine.[33] The enumeration of sixteen scientific and technical fields to be drawn upon by the journal offered a glimpse of the extensive resources now available for the rebuilding of medical theory and practice.[34]

Fourcroy's opinion of physiology was not quite so severe as that of some members of the Comité de Santé, for he perceived it as having begun with Haller's work to turn from mere opinion and speculation and to enter the high road of observation and experiment. Few had yet dared to follow Haller in the way he had opened up. Nevertheless, Fourcroy felt that the science of physiology was now on the point of transforming itself in the wake of recent discoveries in physics and chemistry. These discoveries, joined to the results of comparative anatomy, the natural history of animals and animal experiment, would soon give to physiology a new extension and perfection.[35]

Both in principle and in detail, Fourcroy's views on the relationships of the healing arts to one another and to the sciences gave expression to the program and practices of the old Société Royale de Médecine. Fourcroy had been an active member of the Société almost from its inception, and had been closely involved in its efforts to join chemistry and medicine. A year earlier, in 1790, the Société's *Nouveau plan de constitution pour la médecine en France*, presented to the National Assembly, had called for the reunification of medicine and surgery, fuller integration of all the healing arts including pharmacy and veterinary medicine, and a prominent role for the sciences in all aspects of medical life.[36]

Four years later, Fourcroy's views and those of the by then defunct Société were partially embodied in the program of courses of the newly established Ecole de Santé of Paris. The order of study in anatomy and physiology was to follow a general pattern, first "demonstrating the position, form, and structure of a part," then its use and its varieties in different individuals.[37] The course was to be presented in four sections. An anatomical prolegomenon would consider the nature of anatomy, the general laws of organization, the classification of organized beings and the particular composition of animal substances. This would be followed by a section on skeletology centered on demonstrations of bones, cartilages, and so on, but also including study of related functional problems such as osteogeny, ossification, the mechanical disposition of the articulations, secretions of the marrow, and of the synovia. A third section was to offer a similar presentation of the structure and function of the soft parts: muscles, viscera, blood and lymphatic vessels, nerves, and teguments. The final section would provide a résumé of the course, a survey of comparative anatomy and physiology, and instruction in the technique of anatomical preparations.

The heavily anatomical orientation that is evident even in this brief summary of the course content is striking. The course could be described by Haller's characterization of physiology as *anatomia animata*: it is a course in anatomy filled out or trimmed with discussions of the functions of each anatomical element considered. The teaching is not organized by functions or by phenomena characteristic of the living organism, but by dead anatomy. One major negative consequence is the implicit elimination of all hypothetical physiology, in the sense of physiology pursued apart from empirical study of bodily organization or anatomy. This no doubt reflects—as does the very association of anatomy and physiology in a single chair—the skepticism toward the older forms of hypothetical physiology prevalent in the 1790s, especially among reforming circles, as well as the positive examples of Haller and of the surgical orientation.

Some indication of the extensive resources placed at the disposal of medical theory and practice by the chemical revolution is given in the description of the course in medical chemistry and pharmacy of the *Plan général*.[38] Also revealing is the description of the course on medical physics and hygiene.[39] Under the heading of experiment (*expérience*) applied to the animal economy were to be included accounts of elementary mechanics, acoustics, optics, and hydrostatics as these bore on the animal economy; the properties of gases, of the electrical and magnetic fluids, and of caloric; animal electricity; and the techniques of microscopy and of animal experiment. Here, evidently, was the locale within the new medical faculty of what remained of the older iatromechanism joined with, or translated into, the latest physics. Like the new chemistry, this material would be available for the construction of a new physiology: not as a preestablished system but as a set of analytical tools to be drawn upon in the experimental investigation of the living organism.

Cabanis had given articulate expression to a perception that became widespread in the 1790s: that the older medicine, if not totally bankrupt, was at least open to grave charges of theoretical uncertainty and practical ineffectiveness. For Cabanis and other thoughtful physicians and scientists, the appropriate response to this situation lay not in the abolition of medicine—as some fashionable critics demanded—but rather in its reform. The older medicine had been too audaciously dogmatic in its theoretical claims, too easily and fully seduced by changing scientific fashion. The result was the incessant clash of mutually hostile sects on the one hand, skepticism on the other. Medical practice enjoyed greater stability, but made little progress. A policy of retrenchment was now necessary. The search for first causes of the phenomena of health and illness, the hasty a priori construction of medical systems, the slavish pursuit of scientific fashion—all this must be abandoned in favor of a more cautious empirical approach. Medical theory must become observational, experimental, phenomenalist, analytical. This methodological stance shared some of the same sources as, and was reinforced by, the predominantly prepositivist intellectual climate of France after Thermidor. It would help motivate, and would legitimize, projects of positive reconstruction in a number of areas of medical theory, including physiology, that were initiated in the 1790s and in the following decade. Materials for this task were now available in the accumulated results of various scientific fields, preserved and transmitted to the new generation by the First Class of the Institute, the Muséum, and the faculties and schools of the medical-surgical, pharmacy, and veterinary professions. Physiology, in particular, could now draw on a wide range of well-established knowledge in anatomy, physics, and chemistry, as well as on its own substantial legacy from the seventeenth and eighteenth centuries. The initial response to this situation was to come from the generation beginning its professional training in the 1790s. One of that generation's members was Xavier Bichat.

While serving as aide to Desault during his operations, I several times had occasion to observe these varieties [in the color of the blood] and their relation to respiration. This relationship struck me even before I knew the reason for it. I have since confirmed it by a very great number of experiments on animals.

—*Xavier Bichat* (1800)

3   Bichat's Two Physiologies

Xavier Bichat arrived in Paris at the beginning of July 1794, a twenty-two-year-old surgical student hoping to secure a post as army surgeon after several months of additional study in the capital. Within three days he had attended the surgical clinic of Pierre Desault at the Grand Hospice d'Humanité (Hôtel-Dieu) and had decided to follow Desault's course there. Within two months his exceptional abilities had attracted the attention of Manoury, the second surgeon, and of Desault himself, and within three months he was the favored student of Desault, with a position as *surnuméraire* at the hospital and lodgings with Desault's family.[1] The connection thus formed with Desault, and with the Hôtel-Dieu, was to be decisive for Bichat's career: it fixed him permanently at Paris, provided him with the most advanced surgical training then available, and imbued him with attitudes and skills that would inform all of his future scientific and medical work.

Through the Société Médicale d'Emulation, which he helped found in 1796, Bichat would become one of the leaders of the new generation of medical doctors and scientists beginning their careers in the 1790s. His writings, including four major books in the scant seven years of his career, comprised both well-articulated programmatic statements and concrete research in anatomy and physiology. Those writings would provide a starting point for much of the work of the succeeding generation. In Bichat's career and work both the beginnings of experimental physiology in the France of the 1790s and the convergence of factors that gave rise to those beginnings and shaped their content stand out in clear relief.

In part, these developments were the product of a peculiar confluence of eighteenth-century trends; the eclectic heritage of eighteenth-century physiology, the "vitalism" of Montpellier, the professionalization of surgery and its association with the sciences, and the institution of the private medical course all had roles to play. In part, too, they were the product of circumstances peculiar to the 1790s. In its intellectual dimension the crisis of medicine fostered skepticism, empiricism, and theoretical caution. In its institutional dimension it yielded a new system of medical education with a strongly surgical and practical emphasis. The mood of optimism associated with the movement of reconstruction in France after Thermidor fired the scientific ambition of young physicians and surgeons, and the centering of education and practice in the hospitals opened to them new opportunities for study and research.

Pursuing his scientific career in the midst of these diverse and sometimes divergent pressures and circumstances, Bichat produced not one but two physiologies. The first, essentially classificatory, drew heavily on the legacy of eighteenth-century vitalism. The second, essentially aiming at operative knowledge of function, depended rather on surgical experience and the example of prior experimental investigation from Harvey to Spallanzani. Both would leave their mark on the physiology of the new century.

Bichat's rapid ascent in Paris had had a long preparation in the provinces. The son of a country physician who had taken his M.D. at Montpellier in 1769, Bichat was born in 1771 in the village of Thoirette-en-Bas, Bresse (later the *département* of Jura). The elder Bichat urged his son toward medicine, and no doubt conveyed to him in some form the ideas of Bordeu, Barthez, and other Montpellier professors.[2]

In 1791, after preparatory education at schools in Nantua and Lyons, Bichat began surgical studies under Marc-Antoine Petit at the Hôtel-Dieu of the latter city. Bichat's teachers and fellow students at Lyons were active partisans of the most advanced surgical and medical knowledge then available. The chief surgeon, Marc-Antoine Petit; his first assistant, Jean-Vincent Rey; and at least one of Bichat's fellow students, Louis-Vincent Cartier, had studied with Desault, and Petit had already introduced Desault's teaching methods at Lyons. Rey had also studied internal medicine at Montpellier. It was no accident that Bichat later sought Desault's clinic on his arrival in Paris, and he did not come to Desault unprepared.[3]

With the declaration of war on Austria in April 1792, Bichat was involved in a military hospital in Lyons, and subsequently in the political turmoil of the city. By the summer of the following year the convention was requisitioning physicians and other health personnel, and Bichat was designated for the Army of the Alps. He joined the service of the surgeon Buget at the hospital of Bourg at the beginning of September 1793. In the six months he spent there, Bichat must have gained valuable practical experience, and probably also had time for reading medical literature. He went home in March 1794 to face an uncertain future. The political situation in his region was still troubled. He did not yet feel prepared to practice, and Lyons was in ruins. The most promising course seemed to be to find lodging with an aunt and uncle in Paris, and to complete his training there. After some difficulty in obtaining a certificate of good citizenship from the local authorities, Bichat left for the capital in the last week of June.[4]

The nature of Bichat's studies in Desault's clinical course at the Hôtel-Dieu may be ascertained from a contemporary prospectus describing a typical day's schedule.[5] From six until half past eight in the morning chief surgeon Desault, accompanied by his students, made the rounds of all hospital patients. He selected certain students to keep an exact record of the progress of the illnesses. He examined all patients, pointed out the notable features of each case, and either carried out the necessary dressings himself or had them done by the students in his presence.

The practical surgical lesson, held in the hospital amphitheater from nine to eleven, was the heart of the course. Outpatient consultations were followed by presentation of hospital patients whose treatment had been completed. The central part of the lesson consisted of the operations to be performed that day, with discussion and perhaps a prior demonstration on a cadaver by Desault. Anatomical examination of deceased patients, reports on patients previously operated on, remarks on the character of illnesses then prevalent in the hospital, and an intensive discussion of one disease selected for attention that day rounded out the lesson. Desault required the participation of students at each step.

At three o'clock there was an anatomy lesson, and Desault questioned the students on the previous day's lesson. At half past four the students went to the patients' rooms for dressings and to await Desault's visit at five. At six they reassembled in the amphitheater for further outpatient consultations. Between the practical surgery lesson of the morning and the anatomy lesson at three, and from the evening outpatient consultation until eight o'clock, the students were engaged in practical exercises: making dissections, practicing operations on cadavers, making up equipment, applying bandages. Every tenth day the practical surgery lesson and the anatomy lesson were replaced by a public oral examination in which Desault tested the students on the material taught in the preceding ten days.

Desault's course was thus heavily practical in aim and methods. His goal was to produce competent practicing surgeons by immersing students in the routine of clinical observation and treatment—by giving them daily exposure to surgical technique and human anatomy through the immediate example of the chief surgeon and the most advanced students. Instruction and examination were oral and practical rather than written and theoretical. The student learned by seeing and doing, so that continuity was established between study and practice. For all of this the hospital locale was decisive. The compact association of a steady turnover of a large number of patients presenting a variety of disease conditions, the opportunity for clinical study and regular autopsy of a relatively large number of cases, an amphitheater doubling as a classroom and as operating room separate from the wards, and the ready availability of rooms for dissection and dissection material—these were unthinkable apart from the institutional context of the hospital.[6]

Desault's innovations in teaching methods at the Hôtel-Dieu were of profound importance for the future of French medical and surgical education. His course in clinical surgery, with the course in clinical medicine of his former student Jean Corvisart des Marets at the Charité hospital, was among the very few formal programs of medical and surgical instruction—Antoine Portal's course at the Collège de France was another—to survive the early years of the Revolution. When, at the end of 1794, the laws were passed establishing the new Ecole de Santé, Desault and Corvisart were made professors and their courses, virtually unchanged, were incorporated into the curriculum as external and internal clinics, respectively. Hospital-based medicine, including clinical instruction and extensive cultivation of pathological anatomy, was to be the hallmark of the Paris clinical school of the first half of the nineteenth century, and it is therefore no exaggeration to say that the courses of Desault and Corvisart were the two most important elements of the new école.[7]

Bichat's connection with Desault involved a good deal more than his attendance at the hospital clinical course, important as that was. In return for his lodging with Desault's family he was expected to accompany his teacher on surgical visits in the city and to assist him in various projects. Most significant were the preparation of summaries of the opinions of other authors on subjects that Desault intended to teach, and the gathering of Desault's general ideas, extracts from his lectures, observations, and so on with a view to publication. By early 1795 Bichat had participated in almost every aspect of the work of his teacher, and had attained the unofficial position of *chirurgien-externe* at the Hôtel-Dieu.[8]

Desault died unexpectedly on 1 June 1795 after a brief illness, a victim of overwork and of a profound depression occasioned by the political disturbances of May, which brought to mind his arrest and detention of two years earlier. With the support of Desault's widow, Bichat was able to continue his studies and to finish his teacher's work. He assisted in the publication of a second edition of the joint François Chopart–Desault work *Traité des maladies chirurgicales* (1780), which appeared in October 1795. With the guidance of Corvisart he completed the fourth and final volume of the *Journal de Chirurgie*, adding to the volume a "Notice historique sur Desault" and some of his own first observations. He projected a still more substantial publication, which would give a systematic order to Desault's many observations and insights, and would distill from them the general principles of the *vrai chirurgie* that Desault had wished to create. At the same time, Bichat continued his intensive clinical work and study at the Hôtel-Dieu.[9]

Even while he continued the work of his teacher, however, Bichat was beginning to move beyond it. Paris under the Directory had become the gathering place for an enthusiastic younger generation of medical students and workers, centered around the new Ecole de Santé. Friendships were formed among this group, and students and practitioners began to feel the need for a regular means of association to facilitate mutual instruction and to stimulate by mutual example. Bichat took the initiative in this project. In May 1796 he read the statutes of a new society to his friends Henri Husson and Claude Burdin. With Jean Alibert, Bichat submitted his program to Michel Thouret, director of the Ecole, and after ministerial approval the Société Médicale d'Emulation held its first meeting on 23 June 1796. Among the younger members were André Duméril and Dupuytren, just beginning their careers, and among the elders Pinel, Corvisart, and Cabanis. Corresponding members included Barthez, Bell, Paolo Mascagni, Samuel Soemmerring, and Spallanzani. Bichat contributed a *Discours préliminaire* to the first volume of memoirs, which appeared the following year, and six articles to the second volume in 1798. His own attendance at meetings, however, was often precluded by other, more pressing activities.[10]

In the same month that he was writing the statutes of the Société, Bichat opened his first private course in anatomy. After a brief but substantial drop in numbers and attendance during the Revolution, the private courses in medicine and related sciences had regained the place of importance in French medical education that they had won in the last decades of the Old Regime.[11] Although emphasizing anatomy, Bichat's first course included some discussion of function, and he repeated some of the animal experiments of Haller, John Hunter, Edmund Goodwyn, Galvani, and others. In the fall of 1796 he added a course in operations to his private instruction. When in September 1798 the Directory issued liberalized regulations on halls of dissection and made cadavers officially available to the professors of anatomy, Bichat expanded his anatomy course and himself took charge of separate courses in surgery and physiology. These courses provided Bichat with a small but regular income. More important, they provided a context and an occasion for his theoretical and experimental development of anatomy and physiology as sciences. In this they supplied a much-needed complement to tbe Hôtel-Dieu, where

Desault's course had made practical and clinical interests predominant.[12]

Bichat's development beyond Desault was in no sense a reaction against the work or ideas of his teacher. On the contrary, as he continued Desault's work at the Hôtel-Dieu, he also gave substantial time to the editing of a comprehensive presentation of the principles and practice of Desault's surgery. The first two volumes of this work appeared in February 1798 as *Oeuvres chirurgicales de Desault*, to which was joined a long and appreciative *Eloge* by Bichat. The last part, which included Desault's *Traité des maladies des voies urinaires*, appeared in December.[13]

Although he had become the disciple and collaborator of Desault, Bichat's title of *chirurgien-externe* at the Hôtel-Dieu remained unofficial, and he still lacked a permanent institutional position four years after Desault's death. This became a major concern during the last three years of his life. It was a desire spurred by straitened financial circumstances and parental pressure as well as by his own vaulting ambitions. Bichat's letters to his family reveal that the expenses of books, instruments, and other necessities were a constant threat to his slender budget. "Such is the state of the physician in Paris who wishes to practice with distinction," he wrote to his sister, "that everything is difficulties and weariness in the first years." Such was indeed the reality, though Bichat's parents persisted in suspecting that their son had been seduced by the errant ideas and mode of life of the capital. "I am as anxious not to pass for a *philosophe* as I am ambitious to have the reputation of a distinguished physician," Bichat protested in one letter to his father. Just a little while longer and that reputation would be established, he continued. Then "I do not believe that any physician in Paris will have begun practice with more advantage than I."[14]

When Fragonard, the *chef des travaux anatomiques* at the Ecole, died in April 1799, Bichat joined the ranks of those applying to succeed him. Among his competitors were Duméril and Dupuytren, who had been prosectors at the school since 1795. Probably sensing their advantage, Bichat withdrew from the competition and Duméril was named to the post.[15] As soon as he had published his *Traité des membranes* in December 1799, Bichat sent a copy to the First Class of the Institute, where it was reported on by Jean Hallé. The following February, Bichat's name was presented by the section of Anatomy and Zoology on the list of candidates for the place vacated by the death of Daubenton. Although he placed low in the voting—eleventh on a list of thirteen—it is noteworthy that his candidacy had been advanced solely on the strength of the *Traité*. Bichat subsequently sent copies of his *Recherches physiologiques*, *Anatomie générale*, and the first volume of *Anatomie descriptive* to the First Class within a month of their publication. This reflects a strong desire on his part to gain the approbation of the First Class, and certainly also an ambition for membership in that body. There is no reason to suppose that Bichat would not have attained that goal had he lived.[16]

Some compensation for these initial setbacks at the Ecole and the Institute was provided when, in the fall of 1800, Bichat was named the youngest member of the newly formed Société de l'Ecole de Médecine de Paris. The Société had been founded on the initiative of the government, at least in part to resume some of the functions of the Société Royale de Médecine in relation to public health in the nation as a whole. Members included the twenty-seven professors of the Ecole, the *chef des travaux anatomiques*, and fifteen adjoints from outside the school. Bichat was of the latter group, which included such distinguished names as Cuvier, Jean Baptiste Huzard, Antoine Laurent de Jussieu, and Nicolas Vauquelin.[17] Still, he did not yet have an official title.

The opportunity to obtain one finally came at the institution at which Bichat already had a long record of service: the Hôtel-Dieu. In January 1801 he applied to the Minister of the Interior for the place vacated following the death of Majault, *médecin pensionnaire*. Jean Chaptal, who had already replaced Lucien Bonaparte as minister, was a strong partisan of hospital reform and was eager to staff the Paris hospitals with the best talent available. Bichat had recently published two major works on anatomy and physiology, had several years of experience as Desault's student and as clinician at the Hôtel-Dieu, and was strongly supported by Le Preux and other colleagues there. On 26 January, on the advice of Chaptal, he was named *médecin expectant* by the Commission Administrative des Hospices.[18] Although it simply made permanent and official a situation which had already existed de facto for the preceding five years, this appointment was an exceedingly important one. Hospitals were now at the center of French medicine, both medical practice and the training of new generations. As Ackerknecht has pointed out, a position in the hospital hierarchy was now more important than one in the Ecole.[19] When, therefore, Bichat was edged out by Duméril for the chair of anatomy-physiology at the Ecole the following month, the reverse was not as serious as it might appear. The young surgical student who had arrived in Paris seven years earlier hoping to gain a post as army surgeon was now in a position to exert a major influence on the development of medicine in France.

The expansion of Bichat's interests and activities beyond those of Desault was first reflected in the *Discours préliminaire* that he contributed to the first volume of the *Mémoires* of the Société Médicale d'Emulation. Referring directly to the atmosphere of enthusiasm surrounding the foundation of the Ecole de Santé, Bichat set out the principles which had guided "the young friends of the sciences" in organizing the Société. The major goal of the society would be to associate all the "human sciences," including belles-lettres and morals as well as mathematics, chemistry, physics, and natural history, in the construction of a new medical theory and practice.[20] This was the confident theme of Fourcroy's short-lived journal of a few years earlier, *La médecine eclairée par les sciences physiques*. Bichat, however, was too much the student of Desault and too much immersed in the mood of skepticism prevalent in the later 1790s not to add a word of caution to his program. If the sciences were essential for medicine, excesses were also possible: "We have not forgotten how much the physiologists have invented fictions." The antidote to such excesses would be steady reliance on experience and observation.[21] Again in line with the climate of opinion of the time, Bichat put forward a loosely defined empiricism as the necessary corrective to the past failures of medical theory. Despite these reservations, the program of the *Discours préliminaire* was much more optimistic about the possibilities for medical theory and about the relevance of the sciences for medicine than Desault would have allowed. Without, as yet, giving this program a definite content, Bichat had set himself on a course that would carry him beyond the range of his teacher's work.

The same themes and general program were given much more detailed and explicit development for the particular case of physiology in an undated manuscript, entitled *Discours sur l'étude de la physiologie*, which Bichat prepared as lecture notes for one of his private courses.[22] Here he grappled with fundamental problems of the nature of physiology as a science: the roles of observation of, and reasoning on, physiological phenomena, and the relation of physiology to other sciences. The underlying concern of the essay was to identify the sources of error in physiological investigation and to specify the methods of research and forms of knowledge that would assure for physiology the degree of certainty expected of an established science.

The *Discours* reveals Bichat's intention to build up a science of physiology and his implicit, underlying confidence that this could be done. At the same time Bichat recognized that this task was problematic—that there did not yet exist a secure and clearly defined science of physiology, although elements of such a science existed. This recognition reflects the extent to which Bichat had come to share the negative attitudes toward internal medicine in general, and physiology in particular, that were widespread in medical and scientific circles in the 1790s. Indeed, these attitudes were so strong that anyone undertaking to cultivate physiology was forced to critically examine the very basis of physiology as a science: its subject matter, goals, methodology, and relations to the other sciences. Special attention had to be given to how to go about acquiring reliable, lasting knowledge comparable to that already attained in astronomy, mathematical physics, and chemistry. The atmosphere of the late 1790s established a tension between the urge to build and the consciousness of past failures and uncertainties. Bichat, as the student of Desault, must have experienced that tension in an especially acute form.

Drawing on disparate eighteenth-century sources, he attempted to find a stable middle ground between the skepticism of his teacher and the overconfidence of some of his predecessors. He began to address the problem by distinguishing between a physiology of observation, which was certain, and a physiology of reasoning, which was very often subject to great error. Even within the physiology of observation there were degrees of certainty. The observation of the living organism in health and disease, stressed by the school of Montpellier, was deficient without the rigorous anatomical knowledge stressed by Desault and Haller. Animal experiment might be strengthened by limitation to appropriate kinds of phenomena, as in Spallanzani's studies of digestion, and by the adherence of the experimenter to specified procedural rules. A Newtonian methodological position recalling that of Barthez and

Cabanis, in which speculative hypotheses were criticized, theories became highly general orderings of phenomena, and the search for first causes was renounced, promised to remove the worst excesses of the physiology of reasoning. To complete the task, the results of the successive fashions in physiological explanation occasioned by the physical sciences must be critically sifted. In contrast to his usage in the *Discours préliminaire*, Bichat now termed these "accessory sciences." The final result of this process was to be a science of physiology more modest in its aims, more careful and critical in its methods, and more confident in its results than its predecessors.

The first tangible results of Bichat's scientific efforts appeared in 1798 in the *Mémoires* of the Société Médicale d'Emulation. Volume 2 included six of his articles. Three of these were on surgery and aimed to correct Desault on minor points. In contrast, the remaining three addressed scientific problems. "Sur la membrane synoviale des articulations" announced for the first time in print the discovery of the synovial membranes. "Dissertation sur les membranes, et sur leurs rapports généraux d'organisation" argued that the membranes must be studied apart from the specific organs of which they formed the structural elements, and should be grouped according to their similarities regardless of their different locations in the body. Bichat proposed a threefold general classification of membranes, and noted exceptions to this scheme. This essay was to form the starting point for two of Bichat's major works: the *Traité des membranes*, which appeared in December 1799, and the *Anatomie générale*, first published in August 1801. "Sur les rapports qui existent entre les organes à forme symétrique et sur ceux à forme irrégulière" began the development of the idea of a fundamental distinction between the two lives of the animal organism, the life of relation and the life of organization, which was to figure prominently in Part 1 of the *Recherches physiologiques sur la vie et la mort* of 1800.[23]

In their scientific, even biological ambitions the latter three essays moved Bichat beyond his teacher's primary concern, which was the creation of a rationalized art of surgery. Even so, he continued to draw heavily on his surgical training and hospital position for problems, methods, and observations. His study of the membranes and fluids of the articulations may well have been occasioned by Desault's special interest in luxations, and consequently in the anatomy of the joints. His reliance on considerations of symmetry in distinguishing the two lives of the animal organism might have been suggested by Desault's division of the human body according to planar surfaces. His techniques of anatomical analysis—maceration, insufflation, injection, boiling, and so on— were those he had learned in Desault's dissection rooms at the Hôtel-Dieu. The cadavers anatomized in these studies might have come from autopsies performed at the same institution, though some may also have been procured from other sources for Bichat's private courses in anatomy.

Bichat had constant recourse to clinical and pathological experience gained under Desault or at the Hôtel-Dieu. On one occasion, for example, Bichat appealed to the phenomena characteristic of glands under pathological conditions to show that certain structures around the joints were not glands but extensions of the surrounding connective tissue. Following the suggestion and example of Pinel, Bichat consistently employed the same kind of pathological phenomena as a means to arrive at a general classification of the membranes. Discussing various pathological observations that revealed properties of the serous membranes, Bichat had occasion to touch on another subject of special interest to Desault. "In an aneurism of the heart, I saw the pericardium, which was only capable of giving way a little, partly detached from the portion of the large vessels that it covered."[24] A rich and ever-increasing store of such observations was available to Bichat as he wrote his first scientific memoirs, and they would continue to play a major role in his later, more substantial investigations.

So far as his physiology was concerned, the most important of those later works was his *Recherches physiologiques sur la vie et la mort*, which first appeared in 1800. The *Recherches* is in two parts, a circumstance that reflects not only the duality of the subject matter, life and death, but a more important fundamental duality in approach. The first part develops certain general biological ideas, some of which had been briefly touched upon in the *Discours sur l'étude de la physiologie*; it is the "physiology of reasoning" projected in the *Discours*. In the second part, these general ideas recede into the background in favor of a more narrowly experimental and observational inquiry into the sequence of physiological events following the initial death of the heart, lungs, and brain, respectively; this discussion corresponds to the "physiology of observation" of the *Discours*.

Informing every aspect of Bichat's physiology is his fundamental distinction between animal and organic lives. As he had explained in his *Mémoire* of 1798, this distinction corresponds, on the level of natural history, to the distinction between the plant and animal kingdoms. Within each animal organism it corresponds to the distinction between those organs which are internal, asymmetrical in form, and engaged in assimilative and dissimilative processes, and those organs which are externally directed, symmetrical in form (or symmetrically paired) and engaged in interactions with the external world. Organic life, Bichat held, is continuous in activity, unaffected by habit, and does not depend on a harmony of action between symmetrically paired organs. It is the seat of the passions, and has the heart as central organ. Animal life is intermittent in activity (for example, it is interrupted by sleep), strongly affected by practice and habit, and dependent upon harmony or balance in the action of its constituent organs. It is the seat of the understanding and has the brain as central organ.[25] In his formulation of this distinction Bichat drew on a range of sources, probably including Aristotle, Buffon, Jean Grimaud, Desault, the medical writers of Montpellier, and Cabanis, all of whom are mentioned in the preface to the first edition or in the text.[26]

Also fundamental to Bichat's physiology is his radical distinction between the phenomena of nonliving nature and those peculiar to living organisms, between physics and physiology. Already sketched in the *Discours sur l'étude de la physiologie*, this distinction found its classical expression in Part 1, Article 7 of the *Recherches physiologiques*. "Ceaselessly variable in their intensity, their energy, and their development," the vital laws placed most of the phenomena of the living organism outside the range of quantitative calculation.[27] For Bichat there could never be a relation of identity between physics and physiology, even in principle. He never denied the limited applicability of physics (and chemistry) to certain physiological problems. Nor did such a distinction mean the abandonment of the ambition to raise physiology to the level of the exact sciences. On the contrary, as we shall see, he came to view the vital properties as analogous to gravitation in physics and affinity in chemistry, and to believe that their laws could be established on as firm a basis as, for example, the inverse square law.

The laws would never be the same, however: the primary relation between the sciences remained one of analogy.

The sharp distinction between physical and vital laws recalls the definition of life given at the beginning of the *Recherches*. Life is "the ensemble of functions which resist death." The external world tends toward the destruction of the organism. Inorganic bodies and other living things act ceaselessly on it, and it would soon succumb were it not for the existence of a permanent principle of reaction, which is that of life. Life is thus conceived as a quantity of potential for reaction to the impinging destructive forces of the external world. Like certain forms of impetus in medieval physics, it is self-expending over time. Thus, the difference between the efforts of the external powers and those of the interior resistance, which is the measure of life, is continuously decreasing from birth to death.[28]

A rigorous scrutiny of well-established results of physiological investigation pointed to the existence of two, and only two, vital properties: *sensibility,* or the ability of any part to "feel" that a certain response is called for; and *contractility*, or the ability of the part to respond. That is, every vital response was considered to be a sort of contraction. Further specification of the two fundamental properties in terms of the two orders of life determined a fourfold division of properties, which Bichat represented as follows:[29]

	Genera	Species	Varieties
Vital Properties:	Sensibility	Animal	
		Organic	
	Contractility	Animal	
		Organic	Sensible
			Insensible

Organic sensibility is decentralized: each organ receives and retains its own impressions. It provokes phenomena such as digestion, circulation, secretion, exhalation, absorption, and nutrition, and is common to plants and animals. In the case of animal sensibility, the impressions received are passed on to a common center. It is the basis of conscious sensations, pleasure and pain, and is peculiar to animals and man. Animal contractility is under the influence of the will, is centered in the brain, and is expressed through the action of the voluntary muscles and voice. Each type of sensibility is closely associated with its corresponding form of contractility: animal sensibility with animal contractility, organic sensibility with organic contractility. Nevertheless, animal sensation can occur without giving rise to animal movement, whereas neither organic sensibility nor contractility can occur without the other, in a given organ. As the above tabulation indicates, organic contractility may be further subdivided into sensible and insensible varieties. The former may be observed in the heart, stomach, intestines, bladder, and in general wherever masses of animal fluids are involved. The latter comes into play in the action of the secretory organs on the adjacent blood, the action of each organ on its nourishing juices, and in general wherever the fluids are subdivided into small masses.[30]

The vital properties may be characterized in quasi-quantitative terms. Each individual organism possesses, at any given age, a fixed sum of vital forces. These are variously distributed to the organs in accordance with the peculiar endowments of the individual and with the activities in which he or she is engaged at any given moment. Each organ has a determined sum of sensibility, around which its quantity of sensibility oscillates. The action of each organ is correlated with the fluctuations of this quantity. For example, it is the quantity of sensibility of the blood vessels, and the absorbants, that determines the nature and quantity of the fluids to be transmitted or absorbed.[31] Moreover, the distinctions between organic and animal sensibility and between organic and animal contractility are, according to Bichat, only a matter of degree. Between the extremes, between vision and the unconscious "feeling" of the liver that its activity is required, the difference is clear. The interval separating them is filled, however, at least potentially, by a continuous gradation of organic into animal sensibility. Bichat was careful to point out that such use of the language of quantity did not imply the subordination of vital phenomena to physical laws.[32]

In developing his vital-properties doctrine, Bichat drew extensively on eighteenth-century physiological thought, and especially on the writers of the school of Montpellier. His insistence on the distinction between living and nonliving, between physiological and physical sciences, gave expression to a current of thought that had originated in Stahl's reaction to seventeenth-century iatromechanism and had subsequently become one of the hallmarks of the Montpellier school. Bichat's fidelity to Montpellier is equally reflected in his intention to preserve some role for physics and chemistry in physiological investigation.

From Bordeu, Bichat had the concept of a special principle of life that was imminent in the organs and tissues and ontologically distinct from the conscious mind or soul. Like Bordeu's sensibility, Bichat's vital properties are present in different forms in all living substance (they are not confined to one or two specialized tissues), and their expression may be either "latent" or "manifest." By localizing the vital properties in the elementary tissues, Bichat merely carried one step further the process of differentiation and decentralization of the special principles of life and their attach- ment to concrete anatomical locales that Bordeu had initiated by attributing a distinct form of sensibility to each organ.[33] And Bor- deu, like Bichat, had spoken of sensibility in quasi-quantitative terms: illness is, in some cases, due to excess or deficiency of sensibility in the parts concerned; and variations in human tem- perament are due to differences in the relative energies of the organs. Finally, Bichat adopted unreservedly Bordeu's views on the intimate association of physiology and medicine: continuous reference must be made to the phenomena of the human body in health and disease, and physiological investigation must ultimately inform therapy.[34]

Bordeu had referred all vital phenomena to modifications of a single principle: sensibility. Bichat, as we have seen, invoked two such principles, sensibility and contractility. His starting point for this distinction was undoubtedly Haller's identification of two spe- cifically vital principles, sensibility and irritability. Haller's prin- ciples, like the sensibility of Bordeu, were considered to be immanent in bodily substance and capable of activity independent of the conscious will. They were, however, confined to particular tissues: irritability to muscle, sensibility to nerve. Bichat, striking out on a third way, took over the duality in the vital principle suggested by Haller without giving up the generality asserted by Bordeu. For Bichat, then, there were two vital properties, sensibility and contractility, common—through their different manifestations— to all living tissue.[35]

From Barthez, Bichat had the Newtonian idea, first developed at Montpellier by Sauvages, that the ultimate nature of the principle of life—Bichat's vital properties—could not be the object of scientific investigation. Furthermore, scientific investigation could proceed without such knowledge. Bichat also adopted Barthez's ideas on the quantity and independence of the vital principle: life as a definite quantity of something, different in amount for each individual, self-expanding over time, and not subject to increase or diminution by environmental factors. This notion is, in fact, easier to relate to Barthez's unitary and—to all appearances—immaterial *principe vital* than to Bichat's immanent and anatomically analyzed vital properties. Finally, Barthez, like Bordeu, supported a close reciprocal relationship between physiology and medicine, and thus the application of knowledge concerning the special principle of life to pathology and therapeutics.

In the *Recherches physiologiques* the vital properties of sensibility and contractility, though differentiated into species and varieties, were still assigned to anatomical locales primarily at the level of the organs. In the *Anatomie générale* of 1801, this localization was made more precise by the association of the vital properties with the more finely articulated structural analysis of Bichat's tissue anatomy. In his *Nosographie philosophique* of 1797, Pinel had pointed to the regular association of certain pathological phenomena with discrete membranes, which could, in this sense, be considered in abstraction from the organs of which they were a part. Acting in part on Pinel's insight, Bichat, in his *Dissertation sur les membranes* of 1798, had begun to study the membranes as generalized anatomical elements separable from the organs of which they were the constituents. One kind of membrane could enter into the composition of a number of different organs. The general classification of the membranes into mucous, serous, and fibrous proposed in the *Dissertation* was developed in great detail in the *Traité des membranes* of 1800. The goal of such a classificatory scheme was to expose the fundamental simplicity underlying the apparent complexity of anatomical structure, to reveal a "nature everywhere uniform in its processes, sparing in the means that it employs, prodigious in the effects that it obtains from them, modifying in a thousand ways a few general principles which preside over our economy and which, differently applied to each of its functions, constitute the innumerable phenomena of them."[36]

Such a generalizing and classifying drive was bound to carry Bichat beyond the special case of the membranes. Why should not the human body as a whole be built up out of a relatively small number of elementary components? This, in fact, is the theme of the *Anatomie générale*, in which Bichat introduced his concept of a small number of tissues as the ultimate, irreducible constituents of all bodily structure. As the whole organism is a composite of various organs, so are those organs themselves built of combinations of a small number of homogeneous tissues. "Here we have the real organizing elements of our bodies, whose nature is constantly the same wherever they may be found."[37] To be precise, the human body is composed of twenty-one discrete tissues, which Bichat had isolated and found to be incapable of further reduction.[38]

Aristotle in antiquity, Fallopius in the sixteenth century, and Felice Fontana, André Bonn, and Bordeu in the eighteenth had all noted that particular organs, as well as the body as a whole, were composed of parts that themselves appeared to be homogeneous. Before Bichat, considerable study had been made of particular tissues on both the macroscopic and the microscopic level.[39] Nevertheless, no modern anatomist had yet formulated a program of the generality and scope of that presented in the *Anatomie générale*. In fact the *Anatomie générale* embodies, in transparent form, the expression in anatomy of the program of analysis launched on the philosophical level by Condillac, recently exemplified for the sciences in the chemical revolution of Lavoisier and personally conveyed to Bichat by Pinel's "method of analysis applied to medicine."

Bichat's anatomical program may be termed analytic in both a substantive and a methodological sense. Substantively, it asserts that the elements of the field of phenomena under consideration have been isolated, and that out of these elements, and them alone, the composite phenomena may be reconstructed. The language of bodily form is built from an alphabet of elementary tissues. Methodologically, as Lain Entralgo pointed out, a tissue "is defined by the homogeneity and constancy of its sensory appearance, whatever may be the conditions under which it is observed and the treatment to which it is submitted."[40] Such treatments include dessication, putrefaction, soaking, boiling, baking, and the action of acids and alkalis—methods of analysis that Bichat had learned from Desault and that he could put to ample use in the dissection rooms of the Hôtel-Dieu. If two anatomical elements displayed a uniform appearance and behavior to the senses under all conditions and through all such tests, they were considered identical. Finally, identification of its elementary structural components enabled human anatomy to assume the taxonomic form favored by eighteenth-century thought in general, and by the analytical program in particular.[41]

The *Anatomie générale* completed the physiological doctrine of Part 1 of the *Recherches* by allocating to the vital properties precisely defined seats in the elementary tissues. At the same time, the various elementary tissues were to be distinguished from one another not only on the basis of their physical properties after death but also according to the distribution to them of different kinds and amounts of the vital properties possessed by them during life. It was this aspect of his physiology, intimately united to his analytical tissue anatomy, that Bichat put forward as the basis of a program for the reform of materia medica and for the construction of a pathological anatomy. It was through study of the alterations of the vital properties or precisely specified tissues, in disease and under the action of therapeutic agents, that pathological anatomy and therapeutics would enjoy a new progress.[42]

With this program, Bichat was confident that he had fulfilled the long-standing eighteenth-century ambition to do for physiology and medicine what Newton had done for physics. Physiology would attain the scientific stature of the physical sciences not by a false assimilation to physics or chemistry, but by the establishment of its own analogous but distinctive categories of explanation. The fundamental character of the vital properties as categories—like that of gravity, elasticity, and so on—was based on their irreducibility in nature. "The chaos is only matter without properties: to create the universe God gave it gravity, elasticity, affinity, etc., and in addition a portion had a share of sensibility and contractility." The theoretical generality of the basis provided for physiology and medicine by the vital-properties doctrine emulated the crucial achievement of Newtonian physics: the subordination of a wide range of phenomena to the law of universal gravitation.[43] The brilliance of Newton's achievement had long left physiology, and medicine generally, in the shadows. Now, in possession of a theory of equal generality and power, the physiological sciences could look forward to an equally successful future.

In Part 2 of the *Recherches physiologiques* Bichat turned to study of the physiological mechanisms by which body functions are terminated and death occurs following a sudden and violent event, as in apoplexy, massive hemorrhage, concussion, or asphyxia. Experimental analysis of the sequence of events in such cases was possible because these conditions could be artificially produced on other species, whereas, in general, the same could not be done with most diseases to which man is subject. Moreover—here Bichat's theoretical presuppositions entered in—"the vital laws are so much modified by morbid affections that we can no longer start from the known phenomena of the living animal in order to study those of the dying animal."[44] Experimental analysis was further facilitated by the recognition that every kind of sudden death began with the initial death of one of the three major organs: the heart, lungs, or brain. The deaths of the other two major organs, and of the body as a whole, followed this initial event in a determined sequence that differed for each case. Hence the order of treatment: the sequence of events following the initial deaths of the heart, lungs, and brain, respectively, including the manner of death of the other two major organs and the way in which the death of all the parts derived from the initial events, would be examined in turn. Finally, the results of these analyses would be related to the nature of various diseases which attacked the heart, lungs, or brain.[45] As Michel Foucault pointed out, it is through such an analysis that the physiological fact of death ceased to be a dimensionless interface between life and nonlife, and became a complex, composite succession of bodily events in space and time, a process which would lay bare the interdependences of organic function.[46]

The starting point for these investigations is to be found in Bichat's clinical experience, particularly in his surgical practice at the Hôtel-Dieu, for which the phenomena of sudden death in its several varieties must have been daily fare.[47] Of Bichat's principal sources of information, one—animal experiment—was heavily indebted to operative surgery for its techniques and attitudes, while the other—clinical observation—corresponded directly to the various aspects of Desault's surgical course at the Hôtel-Dieu. Indeed, as Bichat's vital-properties doctrine and general biological ideas evolved out of his critical confrontation with the physiological thought of Haller and the school of Montpellier, so his analysis of the physiological mechanisms of sudden death found its problems and techniques largely in his background as a practical and experimental surgeon.

The surgical provenance of Bichat's experimental approach is well exemplified in his account of the transfusion experiments carried out in the course of his study of the physiological events of asphyxiation. Having just shown that, on the cessation of action of the lung, it was the contact of the dark (deoxygenated) blood on the fleshy fibers of the heart that determined the subsequent arrest of the heart's action, Bichat wished to show that the same held for the brain.

In his initial experiment, Bichat transfused the arterial blood of one dog to the brain of another. He cut the carotid artery of the donor animal, and tied it on the end away from the heart, while adapting a tube to the end toward the heart. He then cut the carotid of a second, recipient animal, inserted the free end of the tube from the donor, and tied it to the end of the artery toward the brain. An assistant, who had meanwhile compressed with his fingers the carotid of the donor animal, released the pressure, allowing the heart of the donor to push its arterial blood to the brain of the recipient. Bichat reported that the recipient animal was not much fatigued by the operation if care was taken to open one of its veins so as to compensate for the gain of transfused blood. He concluded that the contact of foreign red (oxygenated) blood would not alter the cerebral functions of the recipient.[48]

The next step was to try the effect of contact of dark (deoxygenated) blood with the brain. Bichat attached the opened carotid of a recipient animal either to the veins of a donor by a straight tube or to its own jugular vein by means of a tube bent back on itself. Contrary to Bichat's expectations, recipient animals in several such trials remained undisturbed. Finally he noticed that reversed flow in the superior end of the opened artery was more than sufficient to prevent the venous blood, which was under relatively feeble pressure, from reaching the brain of the recipient. To overcome this obstacle Bichat employed a syringe, heated to body temperature. Opening the carotid artery of a single animal and tying the end of the carotid toward the heart, he used the syringe to withdraw dark blood from the jugular and to inject it into the carotid toward the brain. Agitation, quickened respiration, and symptoms of asphyxia were evident almost immediately. The animal life—nervous and motor function—ceased. Heartbeat and circulation continued for half an hour, then stopped, bringing the death of the organic life.[49]

He tried the same experiment on several other dogs, always with identical results: agitation, symptoms of asphyxia, suspension of the animal life, death. Further experiments with a syringe showed that the contact of the dark blood with the air was not responsible for its harmful effects, and that the injection of other fluid substances such as ink, oil, and wine produced the same effects. Exposure of the cerebral surface showed that the brain continued its rising and falling movements even after suspension of the animal life, so that such motion could not be the cause of the latter.

Further experiments showed that asphyxiation could be conveyed from one animal to another via the dark blood. Bichat cut the trachea of a donor animal, then closed it hermetically. After two minutes the blood of the arterial system flowed dark. Opening the carotid, he withdrew some of this dark blood with a syringe and injected it into the carotid of a recipient animal toward the brain. Death of the recipient followed the rapid appearance of the usual symptoms. Similar but slightly different results followed from a variant of the first transfusion experiment. As in that case, Bichat joined donor and recipient dogs by a silver tube connecting their carotid arteries. The trachea of the donor was cut and stoppered in such a way that it could be opened and closed at will. After allowing red blood to flow for a moment, Bichat stoppered the trachea of the donor animal. Dark blood soon flowed from donor to recipient, leading to the expected symptoms of asphyxia. These were not as severe as in those cases in which the syringe was used, however, and if the transfusion was suspended Bichat found that the recipient animal might even survive and live. "Does the air," he wondered, "therefore communicate to the blood some principle still more harmful than that given to it by the elements that render it dark?"[50]

An operator coolly opens a living body, preserving its life during, and in spite of, the intervention. An experienced eye quickly locates major organs and vessels, confident in its unreflective, practical knowledge of their anatomical relations. A skilled hand sections and ligatures blood vessels, opens the trachea so as to control its supply of air, withdraws and injects blood and other fluids with a syringe heated to body temperature, and creates artificial anastomoses of active blood vessels with hollow tubes constructed to bodily dimensions. An assistant compresses and releases blood vessels on the command of the operator. All of these delicate and—to the subject operated on—potentially fatal procedures are carried out as manipulations of organs and tissues at the level of gross anatomy, without reference to putative micromechanical or chemical features of the parts concerned or, for that matter, to their vital properties. All is open, visible, subject to the direct manual or instrumental control of the operator. The procedures and results are recounted in language that is spare, unadorned, lucid, direct: there is no pretence of rhetoric. The cumulative effect is unmistakable: we are watching a surgeon at work.

And yet Bichat is doing physiology, not surgery: his aim is to understand bodily function, in this case the dependence of the brain on the lungs. Practical implications are not overlooked—the investigation leads to the suggestion of a new technique for resuscitation[51]—but they are not the primary end in view. In fact, the means of surgery—its techniques, its instrumentation and a number of its fundamental attitudes—are here placed in the service of a specifically scientific end: the attainment of physiological knowledge.

The subtle but fundamental transition from operative surgery to experimental physiology is evident within Bichat's career itself, and is embodied in his work in experimental surgery. The Faculty of Medicine of Paris possesses a number of surgical manuscripts by Bichat, probably prepared as notes for his private course on "operations." These reveal Bichat's tendency to take an experimental approach to practical surgical problems. Like Hunter and Desault before him, he was interested in the development of the collateral circulation in the limbs following ligature of the major arteries. He devised experiments to test the seeping (*suintement*) of arterial blood that occurs in a limb with the artery so blocked, and suggested a method for creating artificial aneurisms in dogs. These investigations resulted in practical suggestions on the procedures to be adopted in aneurism operations: progressive ligatures so as to allow gradual development of collateral circulation; and—here Bichat cites prior experiments of Desault—elimination of an artery by removal of its surrounding tissue rather than by ligature.[52]

The goal remained the practical one of increasing the effectiveness of operative surgical technique. Nevertheless, such investigation represented a major step beyond a surgery conceived as a mere body of craft skills. The closed, traditionalist, empirical character of the craft gave way to an open-ended, self-consciously progressive process which based surgical technique on experimentally determined knowledge of underlying physiological processes. If this activity was not yet experimental physiology, it would require only an inversion of priorities to make it such: its practitioners had to take as their primary object the physiological processes themselves, and make of the improvement of surgical technique a more or less incidental by-product of the scientific study of nature.[53]

Bichat's experimental physiology was further indebted to surgery for a number of specific instruments and techniques. The eighteenth century had witnessed a steady accumulation of new surgical instruments, including the tourniquet, the forceps, the hernia bandage, the hysterometer, the trepan with crank, tonsils-pinchers, and a variety of lithotomes and gorgets. Among these were items of direct relevance to Bichat's experimental work, such as the surgical syringe and various sounds, bougies, and cannulas made of metal, gum, or rubber. Ligature of blood vessels, such as that employed by Bichat in the experiments described above, had become the procedure of choice for hemostasis only in the last decades of the eighteenth century.[54]

Surgery was not, of course, the only source for Bichat's experimental physiology. Physiology by this time had its own anatomical-experimental tradition descended from Harvey's studies on circulation, a tradition that had itself been a major stimulus to the appearance of experimental surgery. In spite of his occasional strictures on hypothetical physiology—in the manner of Desault and other critics of the 1790s—Bichat was well aware of the achievements of this line of properly physiological investigation, and in the *Recherches physiologiques* he cites its representatives, from Nicolas Steno and Robert Hooke in the seventeenth century to Haller and Spallanzani in the eighteenth. Despite his reservations on the role of the physical sciences in physiology, Bichat was well informed on the recent studies of Lavoisier and other French chemists on respiration, and utilized their results in a qualitative fashion.[55]

Perhaps inspired by Cuvier and Hunter, Bichat carried out a

substantial number of comparative physiological experiments, using as subjects birds, reptiles, and fish as well as mammals. He suggested the connection of nervous and respiratory systems in insects as a subject for future investigation.[56] With a number of comtemporaries, including Humboldt, he made use of the recently discovered connection between galvanism and animal neuromuscular phenomena to clarify the functional relationship of heart and brain.[57]

Finally, Bichat drew upon the publications of the First Class of the Institute for information and perhaps upon its members for encouragement and moral support in his experimental studies. He was familiar, for example, with the commissioners' report on the influence of various asphyxias on galvanism, and with Hallé's study of the symptoms accompanying the mephitism of latrines. The experiments of Part 2 of the *Recherches physiologiques* were witnessed and verified by Duméril and Hallé, and the latter reported favorably on the book for the First Class.[58]

Part 2 of the *Recherches physiologiques*, as the "physiology of observation" projected in the *Discours sur l'étude de la physiologie*, is not restricted to an account of animal experiments and their results. Throughout there is constant reference to clinical-pathological phenomena which complement or supplement the experimental conclusions. The phenomena observed in human subjects suffering asphyxiation are identical to results obtained from this condition as it is experimentally produced in animals. Stoppage of heart action through wounding, burst aneurism, syncope, or polyps serves to clarify the dependence of other functions on the heart. Though other sources, such as Portal, Joseph Lieutaud, Isbrand de Diemerbroek, Giovanni Morgagni and the *Mémoires* of the Académie de Chirurgie are frequently cited, the bulk of the clinical-pathological evidence, including its most vivid and assured instances, is drawn from Bichat's own experience under Desault and at the Hôtel-Dieu. Much of this experience derives directly from the practice of surgical operations. For example, Bichat gives us reason to believe that it was a chance observation during an operation that led to his series of experimental researches on the effects of altered respiration on the blood. Noting that the relative state of the respiration affects the color of the blood in major operations, he recalls that "while serving as aide to Desault during his operations, I had the occasion to observe these varieties several times and their relation with respiration. This relationship struck me even before I knew the reason for it. I have since confirmed it by a very great number of experiments on animals. I have verified it and demonstrated it in the extirpation of a cancerous tumor of the lip, which I carried out last year."[59] If, in the formal argument of the *Recherches*, clinical-pathological experience most often appears as supplementary to the experimental demonstrations, this statement suggests that it also served as the point of departure for at least some of those investigations. Animal experiment and clinical experience were reinforced by observations made on the cadavers of the victims of the guillotine, to which Bichat had access thirty to forty minutes after execution.[60]

In its choice of problems, its skills and techniques, its localism, its phenomenalism, and its use of clinical-pathological material as source of ideas and complementary evidence, the physiology of Part 2 of the *Recherches physiologiques* is heavily indebted to Bichat's clinical surgical training under Desault and to his experience at the Hôtel-Dieu. Yet however much Bichat's experimental physiology drew on that training and experience, it was not a part of it: its appearance required the evolution of Bichat's thought and ambitions beyond the exclusively clinical and practical world of Desault. For Desault's aim was to train effective surgical practitioners, and to perfect the technique and instruments of surgery as a healing art by organizing it according to simplified, rationalized general principles. Even his anatomical studies were harnessed to this end through the emphasis placed on the study of the juxtaposition of organs *in situ*, on functional mechanics, and on pathological anatomy. In addition to these limitations inherent in his positive program, Desault also shared the strongly negative opinion of internal medicine that had become so influential in the 1790s. Bichat refers to this attitude of his teacher in his *Essai sur Desault*:

> Observation was his only guide. Perhaps he was too mistrustful of reasoning in the treatment of disease . . . Is it necessary, from fear of encountering mistakes, never to take the chance of seeking the truth?
>
> Medicine appeared to him to be only an obscure labyrinth, in which we wander at random in the confused paths of error and truth. He cultivated this branch of our art very little, though he borrowed from it his means of healing [that is, drugs]: perhaps he would have been able to unite to a profound knowledge of external

lesions more exact notions on the internal lesions which so often complicate them.[61]

The development of Bichat's program required something more than Desault could provide: a commitment to the possibility of exact knowledge in internal medicine and physiology, a willingness to "take the chance of seeking the truth."

It is, therefore, no accident that it was just at that juncture in his career, when Bichat was beginning to venture beyond the clinical and surgical preoccupations of his teacher, that he began to develop those methodological ideas that figure so prominently in the *Discours sur l'étude de la physiologie* and in the introductory sections of the *Anatomie générale*. The methodological ideas of the *Discours*, in reflecting Bichat's underlying concern to identify the means of acquiring and establishing certain, exact knowledge in physiology, are a means of legitimizing the departure from the straight and narrow way of surgery, an apologia directed at the ghost of Desault and at the skepticism of contemporaries. Those of the *Anatomie générale* serve the additional function of justifying the unifying, generalizing, and rationalizing ambitions of Bichat's program for medicine based on the vital-properties doctrine. The elaboration of a Newtonian methodological position therefore serves Bichat not as a source of the content of his research nor even as a set of rules for its concrete prosecution, but rather as a certification of its character as a scientific enterprise and of its kinship with the other sciences at the most fundamental level.

Recognition that Bichat's physiological program was not a deductive result of his methodological ideas reduces the sense of surprise that might otherwise be felt on discovering its duality. For, as we have seen, Bichat really produced two physiologies, complementary but distinct. One, drawing extensively on the general currents of physiological thought of the eighteenth century, particularly that flowing from Montpellier, took the form of a taxonomy of vital properties and, joining itself to tissue anatomy, aimed to provide a unified basis for pathological anatomy and therapeutics. The other, grounded in clinical and experimental surgery as well as in a tradition of experimental physiology descending from Harvey, took the form of an active surgical intervention in bodily processes which sought to analyze functional organic dependences. This duality of sources, nature, and aims accounts for the sense of hiatus the reader notes between the first and second parts of the *Recherches physiologiques*: there is a disjuncture between the two that goes beyond the difference of subject matter.

Although particular aspects of his physiology have been well characterized by Bichat scholars, this fundamental duality in his approach to the science has gone unrecognized. The consequences for interpretation have been significant. In some accounts Bichat the experimentalist disappears altogether, though the sources and ramifications of his vital-properties doctrine are amply displayed.[62] In others, that doctrine is held to be somehow at the base of all Bichat's work in physiology, anatomy, and pathology, though the precise nature of this relation of dependence is never specified.[63] Still others have offered perceptive studies of Bichat's experimental technique, while scarcely mentioning his vital-properties or tissue doctrines.[64] Few have attempted to present a picture of Bichat's physiological research that comprehends and takes seriously all of its facets.

An exception is the recent study by William Randall Albury.[65] Albury's discussion of Bichat occurs in the context of an attempt to characterize the nature of the changes that ushered in modern experimental physiology in the early decades of the nineteenth century. Like Foucault, he posits the occurrence of a sudden, global transmutation of European thought in the decades centered around 1800. In biology this shift was represented by the transition from the essentially taxonomic enterprise of eighteenth-century natural history, which aimed at the complete tabular grouping of its objects of study according to degrees of identity and difference in a limited number of characters, to study of the organism as a whole in terms of the integrated and adapted structure and function of all its parts, best exemplified in Cuvier's comparative anatomy. In physiology this change took the form of an abandonment of the eighteenth-century taxonomy of vital properties—however named and defined—in favor of the experimental study of functions.[66]

In Albury's analysis Bichat stands clearly on one side of this watershed, François Magendie on the other. One consequence of this posture is that Bichat's physiology must be seen as a unitary entity and must be assimilated to what is held to be the characteristic eighteenth-century form of the science. Bichat's vital-properties doctrine must therefore be seen as the essential basis of his

physiology as a whole. Since, to his credit, Albury is unwilling to
ignore or dismiss Bichat's experimentalism, and yet cannot admit
its autonomy, he is forced into an awkward interpretation of its
place in Bichat's overall program. Thus, logical arguments must
be found to account for the fact that Bichat's notion of the in-
stability of the vital forces had no inhibiting effect on his use of
experiment. An attempt must be made to represent Bichat's use
of observation—both normal and pathological—and experiment
as sectors on a single continuous methodological spectrum, with
observation brought into play for study of the whole and experiment for study of the parts. Although Albury asserts that Bichat
"broke with the traditional concept of function as the unique action
of a single organ," his general thesis impels him to add the apparently contradictory statement that this new conception was
"wholly integrated into his theoretical problematic of vital properties." A similar line of reasoning leads him to identify Bichat's
methods in Part 2 of the *Recherches physiologiques* with those of
Haller in the latter's investigations of sensibility and irritability,
though the use of experiment in the two cases is manifestly different.

Such difficulties are eliminated once the essential duality of
Bichat's physiology is recognized. Bichat's experimentalism is not
inhibited by his vital-properties doctrine because it is largely independent of that doctrine in sources, nature, and aims. The uses
made of observation in the vital-properties and tissue doctrines
on the one hand and in experimental investigations on the other
are distinct, and cannot be seen as coincident points on a single
continuum. The complex real interactions between observational
and experimental findings defy formal codification into stepwise
methodological procedures. The apparent contradiction involved
in Bichat's definition of function dissolves once it is recognized
that the latter applies to his experimental physiology but not to
his tissue doctrine. Finally, the uneasy identification of the use of
experiment in Haller and Bichat is rendered unnecessary by the
admission that Bichat's experimental physiology was not simply a
mapping of the preestablished categories of his vital-properties
doctrine, as was the case in Haller's studies on sensibility and
irritability.[67]

Albury correctly underlines the consonance of Bichat's vital-
properties doctrine and tissue analysis with the eighteenth-century

epistemology of sensations and taxonomic model of order. What ultimately limits his analysis is its exclusive preoccupation with concepts and methods at the expense of context. Bichat's motives for turning to experiment are narrowed to purely methodological and epistemological commitments. Bichat the physiologist appears as a wraithlike intellect floating above the institutional and attitudinal changes of his day, and even above his own professional training, experience, and ambitions. Once defined as pure intellect, the historical actor is expected to construct a science that is logically and methodologically unitary, coherent, and self-consistent. This Bichat did not do. Subject in the Paris of the 1790s to a diverse range of pressures and opportunities, he produced two physiologies that were largely distinct in sources, aims, and methods.[68]

To judge from plans that Bichat had begun to make for a revised second edition of the *Recherches*, his future physiological work would have taken him further in the direction of the experimental approach of Part 2.[69] If so, this would have been only one aspect of what was probably an inevitable centrifugal tendency in his general program. In spite of their abstract unity under vital properties doctrine, each area of medicine with which Bichat was involved—physiology, therapeutics, and pathological anatomy—required its own concrete research approach. And this points to a final paradox in Bichat's physiology: although he aspired to be, through his vital properties doctrine, the Newton not only of physiology but of medicine as a whole, it was the experimental tendency in his work that provided a starting point and exemplar for the scientific physiology of the first decades of the nineteenth century in France.

Without placing the merit of [Bichat's] works in doubt, is it not surprising that today the physiologists of the capital do not dare to undertake anything without having for the text of their research a few passages from his works, a few views that he was able only to indicate? His works have become a species of holy scripture from which one cannot depart without sacrilege. . . . The taste of Bichat for experiments has produced the mania of vivisections, and an unlimited confidence in this manner of studying physiology.

—*John Cross* (1820)

4   A New Generation and a New Program

Bichat's short but prolific scientific career provided a new point of departure for the medical sciences in France. His general anatomy was immediately recognized as a major achievement, and the study of tissue systems became a new field of anatomical research. Bichat's physiology, in its manifestations as vital-properties doctrine and experimental analysis, became a standard point of reference for teaching and for the most important new lines of research. Confirmed or criticized, extended, modified, or rejected, Bichat's writings and teaching provided the French medical world with a glimpse of what physiology might be, of what it might aspire to. In the two decades following Bichat's death physiology began to find its own clearly recognizable voice and position as a science, first in France, then in the rest of Europe.

As Bichat's scientific career must be seen as a product of eighteenth-century trends and of circumstances peculiar to the 1790s, so must the work of his immediate successors be understood by reference to the situation of French medicine and science under the First Empire and the early years of the Restoration. The new system of medical education, with its union of surgery and medicine and its emphasis on practical exercises, technical proficiency, and merit, was confirmed and consolidated by Napoleon. The mood of expansive optimism and ambition among the younger generation of the scientific and medical community likewise carried over into the new century, encouraging dreams of empires different in kind, and more permanent, than those being secured by French armies. Pharmacy and veterinary medicine continued to consolidate their status as liberal professions, and with it their scientific aspirations and activity in fields such as chemistry, anatomy, and physiology. The First Class of the Institute, open since its founding in 1795 to medical professionals and to the medical sciences, offered encouragement and direction to sciences still uncertain in self-definition and weak in institutional support. Membership in the First Class provided an alternative or supplementary focus of ambition for those who might otherwise have confined their careers to the professional faculties, the hospitals, or private practice. The first of the great nineteenth-century voyages of discovery, dispatched to Southeast Asia by the French government in 1800, returned later in the decade with findings that would stimulate the medical sciences as well as botany and zoology.

In this situation a generation of experimental physiologists appeared within the French medical and scientific community. Its members were contemporary with, or slightly younger than Bichat. All were products of the post-Revolutionary system of medical education. All had, at least for a time, scientific ambitions oriented toward the First Class of the Institute. All were more or less subject to the example of Bichat's experimental physiology. Besides Bichat, who may be counted as member as well as exemplar, they included Julien Legallois, Pierre Nysten, Guillaume Dupuytren, and François Magendie.

Not all of their contemporaries with substantial involvement in physiology embraced experimentalism in the image of Bichat. A second group, which included André Duméril, Anthelme Richerand, Etienne Serres, and Henri Dutrochet, adopted other approaches. Richerand, a surgeon and medical essayist, wrote elegant and popular textbooks of physiology in which he presented not his own research but the results of others, including Haller, Bichat, and the Montpellier writers. On occasion he expressed hostility to Bichat.[1] Duméril was named *adjoint* professor of anatomy and physiology at the Paris Ecole de Médecine in 1801 over Bichat and Dupuytren, held that position until 1818, and from 1823 to 1830 held a new, separate chair of physiology at the Faculty of Medicine. His original work was in comparative anatomy, however, and from 1803 he had the chair of herpetology and ichthyology at the Muséum.[2] Serres's research interests followed similar lines. When, in 1820, he was awarded the prize in experimental physiology by the Academy of Science, it was for a two-volume work on the comparative anatomy of the brains of animals.[3] Dutrochet followed medical training in Paris (1802–1806) with a brief career as army doctor. He then settled on his family estate to do research and correspond with the Paris Academy of Science. Inspired by lectures at the Muséum and by the writings of Spallanzani, Dutrochet took up comparative physiology. By the early 1820s he was one of the few French physiologists studying microanatomy and including both plants and animals in his investigations.[4]

By 1820 other experimental physiologists had appeared in Paris. The most important of these latecomers were William Edwards, Félix Savart, and Pierre Flourens. Like the members of the first group, they were products of the new medical education. Also like the members of that group, all had scientific ambitions associated with the Institute. Yet Savart and Flourens were at least twenty years younger than Bichat, and Edwards came so late to Paris, and to physiology, that he must be counted with the younger men. None of the three appears to have taken Bichat's experimental physiology as immediate model. In his science, Savart was a physicist who happened to include the physics of living bodies among his particular objects of study.[5] Flourens was a protégé of Cuvier.[6] Edwards, whose intellectual versatility and independent wealth made possible a varied scientific career, sandwiched his years of physiological research (1817–1824) between prior work in chemistry and subsequent investigations in ethnology and anthropology.[7]

Also active in physiology in the period 1800–1820 were representatives of an earlier generation of the French medical and scientific community. Antoine Portal, François Chaussier, Paul-Joseph Barthez, Jean Hallé, and Charles Dumas figure most prominently among this group. All were from six to twenty-nine years older than Bichat, and all had received their medical education under the Old Regime. Their interest and involvement in science are reflected by the fact that each eventually became a member or correspondent of the Academy of Science. Portal held chairs of anatomy at the Collège de France from 1769 and at the Jardin du Roi from 1778, and in connection with his lectures at those institutions he carried out experiments in physiology and pathology. His work was frequently cited by Bichat, and after 1800 he served on commissions appointed by the Academy to judge the research of the younger experimental physiologists.[8] Chaussier, with Fourcroy, had drafted legislation to create the new Ecole de Santé in 1794, and until 1822 he served as the school's first professor of anatomy and physiology. His major publication, however, was an attempt to provide a systematic classification and nomenclature for the muscles of the human body; and unpublished manuscript notes to his physiology course of 1806–1807 reflect little awareness of recent work or evidence of original research in physiology.[9] Barthez carried the Montpellier tradition into the nineteenth century with a second edition of his *Nouveau éléments de la science de l'homme*, published in 1806. Though a product of the Old Regime, Hallé consistently allied himself with modern science and with progressive trends in medicine. An early member of the Société Royale de Médecine and of the Société Médicale d'Emulation, he was named first professor of hygiene and medical physics at the Ecole de Santé, and professor of medicine at the Collège de France in 1804. As member of the First Class of the Institute he frequently encouraged work of the younger generation in experimental physiology, and he collaborated directly with Bichat's student, Pierre Nysten, in the writing of articles on hygiene and medical physics for the *Dictionnaire des sciences médicales*.[10] Dumas, a professor of medicine at the Medical University of Montpellier, was named professor of anatomy and physiology at the new Ecole de Santé of that city in 1795. Like Chaussier, he published works on the systematics of human musculature, and, in addition, a textbook of physiology that presented a variant of the eighteenth-century vital-properties doctrine.[11]

Analysis of the population of those involved in the research or teaching of physiology in France from 1800 to 1820 therefore yields four groups: an older generation, holdovers from the Old Regime; a younger generation, trained in the new system of medical education in the years immediately following its establishment, and divided into those who pursued original experimental research and those who adopted other approaches; and a group of latecomers, experimentalists who did not begin original research in physiology until the second half of the 1810s (see Tables 1 and 2). Of these groups, the experimentalists of Bichat's generation are of special interest, for it was their version of physiology that had begun to prevail in the French scientific world by the 1820s and that would be the principal model for later generations.

Belgian-born, in the same year as Bichat, Pierre Nysten came to Paris in 1794 and entered the first class of the Ecole de Santé.[12] Distinction in his studies won him a place in the first class of the revived Ecole Pratique, and in 1798 a position as aide in anatomy in the medical school. The first decade of the century he devoted to experimental physiology and to various government medical

Table 1. Individuals teaching or doing research in physiology in France, 1800–1820, arranged by order of date of birth.

Name	Dates
Paul-Joseph Barthez	1734–1806
Antoine Portal	1742–1832
François Chaussier	1746–1828
Jean Noël Hallé	1754–1822
Charles-Louis Dumas	1765–1813
Julien Legallois	1770–1814
Xavier Bichat	1771–1802
Pierre Nysten	1771–1818
André Duméril	1774–1860
Henri Dutrochet	1776–1847
William Edwards	1776–1842
Guillaume Dupuytren	1777–1835
Anthelme Richerand	1779–1840
François Magendie	1783–1855
Etienne Serres	1786–1868
Félix Savart	1791–1841
Pierre Flourens	1794–1867

Table 2. Grouping of individuals listed in Table 1.

Holdovers	Experimentalists of Bichat's generation	Others of Bichat's generation	Late-comers
Portal	Legallois	Duméril	Edwards
Barthez	Bichat	Richerand	Savart
Chaussier	Nysten	Serres	Flourens
Hallé	Dupuytren	Dutrochet	
Dumas	Magendie		

assignments. His chosen science did not yet offer a career, however, and after 1810 he turned to private medical practice and away from physiology. Though never a member of the Institute, he was a collaborator and favorite of Hallé, who helped him obtain a position as physician at the Hospice des Enfants.

Nysten's experimental physiology was directly and explicitly modeled on Bichat's. Shortly before Bichat's death the two men had agreed to collaborate on experiments on pathological physiology, meaning the experimental study of the alterations of functions in disease. They expected these studies to replace Part 1 of the *Recherches physiologiques* in a new edition.[13] In the introduction to his first published research, on the degrees of contractility of various muscles when subjected to galvanism, Nysten referred to Bichat's experiments on the effects of galvanism on the hearts of warm- and cold-blooded animals, and gave a long quotation from Part 2 of the *Recherches physiologiques*.[14] The text was divided largely on the basis of whether death was due to the initial death of the heart, lungs, or brain, a clear extension of Bichat's approach. In his investigations on physiology and chemical pathology, undertaken "to extend those of Bichat on life and death" and published in 1811, Nysten made a systematic study of the effects of the introduction of various gases into the venous system, showing that death did not inevitably follow such a procedure but that the exact sequence of events depended on the type and quantity of gas introduced.[15]

Six years the junior of Bichat and Nysten, Guillaume Dupuytren began his medical studies in Paris in 1793, at the age of sixteen.[16] He studied anatomy with the surgeon Alexis Boyer, chemistry with Edme Bouillon-Lagrange and Vauquelin. One of the first students at the Ecole de Santé, he competed for and won a place as prosector in 1795. Perhaps with the encouragement of Chaussier and Claude Leclerc (the adjoint professor of anatomy and physiology), he began to do physiological experiments, and supplemented his income by offering private courses in physiology and anatomy. With Bichat, he was a founding member of the Société Médicale d'Emulation in 1796. Talented and fiercely ambitious, he worked long hours in spartan living conditions, much in the manner of the young surgeon Desplein in Balzac's "The Atheist's Mass." In 1799 his reputation was already such as to move the medical school to obtain his exemption from military service so that he could continue his work as prosector and his research in physiology.

By the time he took his M.D. in 1800, Dupuytren was poised on the threshold of one of the great careers of his generation. Over the next fifteen years he advanced from step to step, post to post, simultaneously in the hierarchies of the medical school and of the Hôtel-Dieu. *Chef des travaux anatomiques* at the Ecole de Médecine in 1801, he was named professor of operative medicine at the renamed Faculty of Medicine in 1812, and finally exchanged this chair for that of clinical surgery in 1815. At the Hôtel-Dieu he was appointed surgeon second class in 1802, *chirurgien-adjoint* in 1808, and chief surgeon in 1815.

With each advance in his career, Dupuytren was drawn further into surgical practice and away from physiological research. The change was gradual. From about 1803 he taught a course in physiology at the private Collège des Etudiants en Médecine, and he was carrying out physiological experiments at Alfort with the director of the veterinary school, Jean Dupuy, well into the first decade of the century.[17] Yet his interest was increasingly drawn by surgery and pathological anatomy, and at this time he began the collection of specimens that would later form the basis of the Musée Dupuytren. By 1810 he had left experimental physiology behind. When in 1825 he was elected to membership in the Academy of Science, it was only over the objections of some members, including Geoffroy Saint-Hilaire, who wished to exclude surgeons who might too easily neglect science for practice.[18]

The experimental physiology cultivated by Dupuytren in the

early years of his career was in the pattern of Bichat's, and the sort encouraged by the First Class of the Institute. Movements of the brain, modifications of substances in the alimentary tract, effects of ligature of the thoracic canal on nutrition, and the roles of the nerves in respiration were among his topics. The latter research, reported to the Institute in 1807, may serve as example.[19] Dupuytren began with the recognition that Bichat's experiments on the dependence of the lungs on the brain via the brain's control of the muscles of respiration did not fully resolve the issue. Comparative structural or analogical reasoning alone could not do so, and further experiments were required. Working at Alfort with Jean Dupuy, Dupuytren explored the effects of sectioning nerves supplying the lungs in horses and dogs. He concluded that the nerves played an essential role in the changes effected in the blood during respiration, and that inspiration and expiration alone were not sufficient to explain those changes.

While such a conclusion was implicitly critical of Bichat on a particular point, the aims and form of the investigation were entirely in the pattern of Bichat's experimental work. The goal was the study of functional dependences through surgical intervention in the living organism. The approach was analytical in that the problem was isolated and did not depend on a holistic or systematic concept of the organism. Operative skill and practical knowledge of gross anatomy were of central importance, and vital properties were not mentioned. Pinel and Hallé reported to the First Class that Dupuytren's experiments "may be placed among those of the first rank in physiology," and that he should be encouraged to continue his work. Dupuytren had proposed to carry out new experiments with the chemist Louis Thenard on the changes of the air and blood under the different conditions described in the 1807 memoir.[20] The sequel never appeared, however, for by that time even heady promises of future scientific achievement were yielding before an unfolding surgical career.

A still more promising beginning in experimental physiology was made by Julien Legallois, Bichat's senior by one year.[21] His early studies at the Faculty of Medicine of Caen cut short by the Revolution, Legallois made his way to Paris to become a student in the clinics of the Paris hospitals. A member of the first class of the Ecole de Santé, he took his M.D. in 1801 with a thesis on the changes undergone by arterial blood as it passes through the organs. That work revealed to him the potential of experimental physiology and his own aptitude for research; and in spite of poor eyesight and meager financial resources, much of his energy and attention for the remainder of his short career was devoted to science. He made a living as physician to the poor of Paris's twelfth district, and a year before his early death in 1814 he was appointed physician to the Bicetre hospital.

An original thinker who used critically a variety of sources, Legallois was by no means simply an emulator of Bichat. In fundamental aims and methods, however, his physiology did not differ significantly from that of the author of Part 2 of the *Recherches physiologiques*. In the investigations undertaken after his thesis, Legallois sought above all to determine experimentally the conditions essential to the life of the organism, in part or as a whole. His most important work explored the dependence of bodily function on the nervous system. Against many of the advocates of the vital-properties doctrine, including Bichat, he demonstrated that the source of sensation and movement was localized in the spinal cord. Maintenance of life in a part depended on two factors: the integrity of the corresponding part of the cord, and a continuous supply of arterial or oxygenated blood. Death could be partial or complete, depending on the state of the cord and the circulation. Against Haller and his school, he showed that the movement of the heart depended not merely on its own muscular irritability but also on the nerves and spinal cord. One set of experimental studies determined the sequence of events leading to death by asphyxiation following sectioning of the eighth pair of nerves; another showed that respiratory movements are controlled by part of the medulla oblongata.[22]

Though informed by a common purpose, each of Legallois's investigations could stand on its own merits, and in this sense his approach to physiology was analytical. A consummately skillful operator, he relied heavily on techniques of nerve section and ligature. Aware of the comparative studies of Cuvier, he made use of reptiles as well as warm-blooded animals as experimental subjects. His work was enthusiastically received by the First Class. Reporting on Legallois's memoir on the dependence of the heart on the spinal cord, Percy, Humboldt, and Hallé praised him for bringing to physiology the "precision and severe logic" of the physical sciences, and for producing a rigorously interconnected and interdependent series of experiments.[23] It was Legallois who, of his generation, best expressed the centrality and aims of animal experiment in contemporary physiology:

> Experiments on living animals are among the greatest lights of physiology. There is an infinity between the dead animal and the most feebly living animal . . . It does not suffice to observe the simultaneous play of all the functions in the healthy animal; it is above all important to study the effects of the disturbance or cessation of such and such a function. It is in the determination by this analysis of the function of such and such an organ and its correlation with the other functions that all the art of experiments on living animals consists.[24]

Fatal illness brought a premature end to the endeavor for Legallois, as it had for Bichat and as career pressures had done for Nysten and Dupuytren. The field was left to the last member of the group, François Magendie.

Among those of his generation, Magendie deserves special attention. It was he who most explicitly and forcefully articulated a definition and goals for experimental physiology as a science. He was the most prolific of concrete research in the period 1800–1820. It was his work that was first recognized by the Academy of Science through election to membership, and by 1821 his leadership in the field was acknowledged by most of the French scientific community. Of that early generation, he had the most sustained career commitment to experimental physiology.

Like Bichat, Nysten, Dupuytren, and Legallois, Magendie was the product of political, institutional, and intellectual circumstances peculiar to France between the Revolution and the First Empire.[25] Growing up in the unsettled conditions of the Revolutionary period and its aftermath, the son of a politically involved surgeon with Rousseauist views on education, Magendie developed from an early age the energy, directness, independence of judgment, and contempt for convention that were to mark his scientific as well as his personal style. He began his medical training in 1799 under Boyer, a surgeon and student of Desault, and his formal medical studies in 1801 at the Ecole de Médecine of Paris. Magendie was thus among the first to be trained entirely under the new system of medical education, with its emphasis on clinical experience in the hospital and practical exercises for students at

all levels. Paternal encouragement, the nature of his first apprenticeship, and the practical emphasis of his education combined to
direct his initial activity and ambitions toward anatomy and surgery. Appointed aide in anatomy at the Faculty of Medicine in
1807, he was already teaching anatomy and physiology by the time
he passed his final examination for the M.D. in 1808. His teaching
continued through his appointment as prosector in 1811 and for
a further two years in this post. In the same period he gave well-
attended courses on operatory surgery at the Ecole Pratique, and
his surgical skills were evidenced by his invention of a new procedure for resection of the lower jaw.[26] Yet his first important
publications, dating from 1809, were on physiology and drug action, and these areas finally won his full attention.

The turn toward physiology was facilitated in a negative way by
the closure of the path to a career at the Faculty. Magendie faced
personal friction with the professor of anatomy, Chaussier, and
perhaps also the professional jealousy of the professor of surgery,
Dupuytren. In 1813 he resigned his post at the Faculty, opened
an office as a practicing physician, and organized a private course
in physiology. This course enjoyed a great success.[27] In 1816–
1817 the first edition of the *Précis élémentaire de physiologie* was
published, partly in response to the interest evoked by the
course.This very influential work was to go through five French
editions and be translated into English, German, and other lan-
guages.

In 1818 Magendie was named to the Bureau Central des Hospices Parisiens as a result of a competitive examination and recommendations from the Academy of Science. He had no official
hospital assignment until 1826, however, and until then had to
rely on his friend Henri Husson in order to observe treatments
and to give a clinical course at the Hôtel-Dieu. An official hospital
position materialized only in 1830, with Magendie's appointment
as director of the women's ward at the Hôtel-Dieu.

Magendie's shift from surgical and anatomical to physiological
interests was paralleled on the institutional level by a primary
orientation toward the First Class of the Institute. At the Institute
he read his first published memoirs, and membership in the First
Class came to be his highest ambition. From an early date he won
the interest and respect of such figures as Cuvier and Pinel, and
the active patronage of Laplace. After eight years of private teach-

ing and publication following resignation of his post as prosector at the Faculty, he was elected to membership at the Institute in 1821. From this point on, the future of Magendie's career, and of his conception of physiology, was assured. He became the representative of physiology within the Academy, its personal embodiment—there was no separate section of physiology—and champion. Members of the medical profession aspiring to membership in the Academy henceforth had to win his support, and his judgments on contemporary research and writing took on new authority.

The year 1821 saw, as well, the publication of the first edition of Magendie's *Formulaire*, the culmination of twelve years of experimental and clinical study of the effects of drugs on the human or animal organism.[28] These studies had been initiated in 1809 with the administration of drugs in the form of crude plant products. With advances in analytical chemistry between 1807 and 1820, and particularly with the discovery of plant alkaloids through the work of the pharmacists Pelletier and Caventou, the active principles of drugs in their pure form were for the first time made available for experimental and clinical trials. Magendie followed these developments closely, and collaborated with Pelletier in the isolation of emetine. The *Formulaire* provided physicians, for the first time, with a list of drugs whose dosage could be determined with quantitative precision and whose effects on the organism had been carefully tested in animal experiments and in clinical trials. Between 1821 and 1836 the *Formulaire* passed through nine editions, undergoing considerable expansion. It became the starting point for modern experimental pharmacology in both its clinical and its physiological dimensions.

Magendie in 1821 also founded the *Journal de physiologie expérimentale et pathologique*, which he personally edited for the next ten years. The *Journal*, the first of its kind in Europe, provided him with yet another means of promoting his own conception of physiology as a science. It served as a forum for his own views at a time when he had no official teaching position, and he controlled the selection of articles for publication. The connection of the *Journal* with Magendie's need for a forum is reflected in the timing of its end. Publication lapsed in 1831, within weeks of Magendie's appointment to succeed Joseph Récamier in the chair of medicine at the Collège de France. For the balance of his career, Magendie's activities moved between the three centers of the Hôtel-Dieu, the Collège, and the Institute. The Hôtel-Dieu provided a steady stream of clinical and pathological experience, and opportunities for the testing of new medicines. At the Collège he carried out the research that created the content of physiology, lectured on his most recent results to medical and scientific audiences, and trained a number of select students—among them Claude Bernard—in the techniques of research. At the Institute he represented physiology to the wider scientific community, judged the work of others in the field, and served on a number of special commissions.

The completion of a first series of researches on absorption and drug action, the publication of the *Formulaire*, the election to the Institute, and the founding of the *Journal* mark the year 1821 as both a culmination and a fresh point of departure for Magendie's career. The pattern of his scientific style and the direction of his research interests were largely determined. Major achievements in physiology still lay ahead, but they would occur within a pattern that had already been set. The primary task of the future would be the institutionalization of this pattern, of physiology as Magendie conceived and practiced it. It is in the dozen years between 1809 and 1821 that we must seek the origins of his physiology and identify the conditions that made it possible.

Magendie's scientific career began with the nearly simultaneous publication of his first experimental work on drug action and absorption and of a critical program for the reform of physiology. Since general programmatic statements are not often found in Magendie's later writings, this early article of 1809 is worth special attention.[29]

Physiology, Magendie pointed out, contrasted unfavorably with the physical sciences such as astronomy in the degree to which its phenomena had been organized under a small number of general principles. A science could not be a mere aggregation of disconnected facts. The "general and essential" phenomena of organized beings must be sought. His own candidates for this role were nutrition, or the series of invisible movements by which living bodies decomposed and recomposed themselves; and action, or those movements, visible or invisible, by which living parts accomplished certain ends. Examples of the latter included such diverse phenomena as the contraction of voluntary muscle and the secretion of bile by the liver.

The general and essential phenomena of physiology in turn required a unitary principle of explanation. This principle would be, like planetary and molecular attraction, unknown in its nature but manifest in its effects. Among the many names given it, Magendie considered the best to be "vital force." The expressions of the vital force would be determined by structure, for "the nature, disposition, and mode of union of living molecules—in a word, the organization—modifies the vital force in such a manner that the phenomena by which they are manifested in the living body are always in direct proportion to the organization." Physiological explanation therefore involved specification of the way in which the phenomena manifested by a given part were determined by its organization. Ideally, such specification would occur at the molecular level. Since this was not accessible to current means of investigation, however, Magendie recommended the example of the zoologists (Cuvier was clearly meant) in determining the relation between organization and function at the level of gross anatomy.

Magendie's notion of physiological explanation based on a unitary vital force and on organization as the principle of differentiation included a twofold critique of contemporary vital-properties doctrine as it was advanced by Bichat and others. The so-called vital properties were not, in the first place, notions directly based on observation. Instead, they interposed needless explanatory entities between the vital force and its expression in phenomena. These entities lacked generality. They could not be common to all living bodies in the same way that extension, divisibility, impenetrability, and mobility are general properties of all bodies. If they continued to assign a special vital property to each particular phenomenon, physiologists would find themselves multiplying vital properties to infinity and would end in a jungle of verbiage. Better to study the visible phenomena as such, and to seek out the relations among them.

With this, Magendie's critique of vital-properties doctrines led into his positive conception of function. If, for example, one did not consider animal sensibility as a vital property, how were the phenomena collected under this heading to be characterized? Simply as the collective end of the actions of a certain number of organs—that is, as a function. The eye receives an impression; the optic nerve transmits it; the brain perceives it; the sum of these three actions is *sensation*. Digestion is the general result of the action of the digestive organs, circulation the general result of the action of the circulatory organs. Such a conception of function provided a way of further differentiating physiological phenomena below the general categories of nutrition and action. Moreover, by dissolving the vital properties of animal sensibility and contractility, it cleared the way for a unitary explanation of the phenomena of life.

Magendie's concept of function also pointed to a third basic concern of his discussion: the mode of investigation in physiology. In the beginning of the essay, he had already referred to the need to confirm most "physiological facts" by new experiments, as one condition of the future progress of physiology. By the conclusion, the rationale behind the need had been more clearly defined. The ultimate nature of the vital force was unknown and probably unknowable, and organization at the molecular level was inaccessible to currently available tools of investigation. Therefore, "it would, perhaps, be as advantageous to begin the study of physiology at the instant when the phenomena of living things become appreciable to our senses."

Combining the basic elements of grouping, explanation, and mode of investigation, Magendie's general program for physiology may be compactly characterized. It is the experimental investigation of functions at the level of phenomena accessible to the senses in order to determine the laws through which the vital force acts, and specifically how these laws are determined by the anatomical organization of the parts.

When he published "Quelques idées" in 1809, Magendie had recently completed his formal medical studies. His background and talents inclined him toward surgery, but he had also been teaching anatomy and physiology successfully for almost two years. He had made himself familiar with the current literature of French physiology—Bichat, Chaussier, Barthez, Richerand, and Dumas are mentioned or alluded to indirectly in this essay—and was aware in at least a general way of the recent work of the First Class of the Institute, particularly in zoology. The direction of his future scientific work and the attitudes that would underlie it were not yet settled. As he prepared his first major research on drug action and absorption for publication in the same year, he must have felt a need to give a clear definition to his position and to justify to himself and to his medical and scientific colleagues the approach he had adopted.

Such definition and justification were particularly called for when physiology was chosen as the field of investigation. The negative attitudes toward physiology that characterized the 1790s were by no means dissipated, either within medical circles where the more narrowly clinical attitudes of Desault and Corvisart were increasingly prevalent or within the scientific community. Despite the work of Bichat, the status of physiology as a science was still questionable; and as a field it lacked an authoritative institutional base, within the Institute or elsewhere.

"Quelques idées" gives expression to Magendie's personal dissatisfaction with the contemporary state of physiology as a science and his sense that it stood in need of a conceptual and methodological reorientation. Although Bichat's ideas are a principal target of criticism, Magendie's project to construct a new science of physiology and his appeal to the model of the physical sciences echo Bichat's stance in the *Recherches physiologiques*, as well as that of Barthez in his *Nouveaux élémens de la science de l'homme* (1778).[30] The desire to align physiology with the physical sciences and the praise of Cuvier's zoology also clearly reflect Magendie's personal orientation toward the First Class of the Institute, his identification with its values and standards, and his desire to gain the approbation of its leading members.

There is little that is novel in Magendie's remarks on the vital force, particularly if these are seen against the background of physiological thought at Montpellier in the eighteenth century. Magendie's "vital force" is simply another variant on this tradition, perhaps closest to Barthez's *principe vital* in Magendie's insistence on its unitary character and on the need to discover its laws through study of its sensible manifestations.

Magendie may have learned of Barthez's views from Cuvier. In a report to the First Class in 1806, Cuvier had commented favorably on the Newtonian analogy for physiology presented by Barthez "in his celebrated work on the *Elements of the Science of Man*, which brought about in its day a fortunate revolution in physiology."[31] Cuvier emphasized that the abandonment of the search for first causes and the naming of the vital principle were not enough, however. The phenomena and laws to which the vital principle gave rise would also need to be studied, for "only thus can physiology flatter itself with having a particular principle in the same way that astronomy has one." The means to these laws would be anatomy, the chemistry of organized bodies, observation of the organism in health and disease, and comparative studies of all classes of living bodies. The parallels between this discussion and Magendie's argument of three years later are striking, and probably not coincidental.

Although Magendie cites Cuvier's holistic concept of body "organization," his own use of the term never departs noticeably from the less specialized meaning of ordinary anatomical structure at all levels. In employing "organization" as the means by which the effect of the unitary vital force is differentiated in particular parts, Magendie employs a sense of the term already in current usage—for example, in Bichat's *Traité sur les membranes.*[32]

It is in his critique of Bichat's doctrine of vital properties and his advocacy of a functional approach to physiological phenomena that Magendie appears most clearly as innovator. Even here, however, his originality must be qualified, since his rejection of vital properties in Bichat's sense is largely a consequence of his appropriation of attitudes then prevalent among zoologists, who had localized sensibility and contractility in the neuromuscular system.[33] Moreover, the cogency of Magendie's criticism is somewhat vitiated by his failure to grasp the fundamental character of Bichat's vital-properties doctrine. Of his two objections, one—that the vital properties are not observable—also applies to Magendie's own vital force.[34] The other—that at least two of the vital properties do not underlie all vital phenomena—misses the point that for Bichat there were ultimately only two vital properties, present with variations in all tissues.[35]

Much more seriously, however, Magendie fails to grasp the basic duality of Bichat's approach to physiology, which comprised both a taxonomy of vital properties and the experimental investigation of functional organic dependences. Detached from the classificatory structure in which they originated, the vital properties are bound to appear superfluous or, at best, puzzling. If physiology is concerned not with the sorting of bodily phenomena into genera, species, and varieties but with the dependence of functions on organization and on one another, then another approach is indeed required. Not the establishment of adequate general categories, but the operative manipulation and control of bodily events becomes the essential desideratum. Magendie so thoroughly opted for this second approach—in fact, there was no question of conscious choice—that the first was placed beyond comprehension. It is doubtful whether Magendie's conception of function ("the common end of the actions of a number of organs") added anything of significance to that already put forward by Bichat, and perhaps by others.[36] In any case, such a definition of function had been implicit in concrete physiological research since the sixteenth century, and its explicit formulation by Bichat and Magendie represented a self-conscious affirmation of the value of the functional approach, rather than its initiation. In practice, Magendie was to be heavily indebted to Bichat's functional, experimental physiology as represented in Part 2 of the *Recherches*. In rejecting one aspect of Bichat's physiology Magendie would, in fact, make himself the leading heir of the other.

Magendie's debts to Bichat were already visible in the thesis he presented for his M.D. in 1808.[37] To an examining committee that included the surgeons Alexis Boyer and Anthelme Richerand and the professor of anatomy and physiology at the Ecole de Médecine, Duméril, he presented a short study of the uses of the soft palate, followed by several propositions on the fracture of the rib cartilage. The latter were Hippocratic aphorisms, compact and pithy, for the benefit of the busy clinician. In contrast, the aim of the discussion of the palate was physiological. The structure and anatomical connections of the part were explored to elucidate its function. Methods were restricted to observation of the living human subject, anatomical study of cadavers, and perhaps observations on himself. Study of alterations of the palate in venereal and other conditions had helped to clarify its physiology. Echoing Barthez, he advised that "it is in vain that the attempt is made to separate the physiology of the healthy man from the physiology of the sick man." The only animal experiments cited were those quoted verbatim from Bichat in support of Magendie's argument on the movement of the palate in inspiration and expiration. Bichat's writings were cited on two other points, once in criticism, once in commendation. In each case an anatomical and functional point was at issue, and nowhere was mention made of the vital properties.[38]

In 1809 Magendie published two other memoirs in addition to "Quelques idées," both accounts of concrete investigations devoid of any programmatic reference. The first, carried out in collaboration with the botanist Rafenau Delille, was a study of the physiological mode of action of *upas tieuté* and related members of the *strychnos* family, from which strychnine was later isolated. The second, occasioned at least in part by observations made in the course of the drug action experiments, was a study of the manner in which rapidly acting drugs such as *upas* were taken up by the circulatory system for transportation to a specific site. Both studies addressed fundamental physiological problems with important implications for medical practice, and each was to serve as point of departure for a series of investigations that would extend over much of Magendie's career. It is primarily in these concrete investigations, and not in his programmatic statements, that we must seek significant insight into the background, character, and institutional circumstances of Magendie's physiology.

Henceforth I became, so to speak, master of a phenomenon which until then had been an impenetrable mystery for me. Because I was able to oppose its development, produce it, make it fast, slow, intense, weak, its nature could not easily escape my investigation.

—*François Magendie* (1820)

5 The Experimentalist in Action

The decade that opened with Magendie's memoirs on drug action and absorption proved to be formative for his science. By its end, his physiology had acquired the essential characteristics that would mark it for the rest of his career. Closely linked with anatomy, its primary goal was to elucidate functional interdependences through direct surgical intervention in the living organism. It was analytical in that each investigation aimed to isolate the structural elements on which a given function depended, and in that the results of each such investigation could stand alone, without reference to some more general system of physiological explanation. The use of chemistry and physics as tool sciences remained an important desideratum. Underlying the whole endeavor was a commitment to attain what Claude Bernard would later assert as the ultimate goal of experimental physiology, both as knowledge and as power: the operative control of the phenomena of the living organism.

The First Class of the Institute played a decisive role in the formation of this physiology through its unremitting emphasis on the centrality of experiment, and through the desire of its leading members that experiment be complemented by comparative, physical, and chemical approaches. Within the First Class, the scientific maturation of physiology was signaled by the standardization and official recognition of the terms "physiology" and "physiologist," and by the founding of an annual prize.

Experimentalism, in the form given it by Magendie and sanctioned by the Institute, took the leading position in physiological research. This was the crucial step that constituted modern physiology in the first two decades of the nineteenth century. Earlier use of experiment in the study of animal—and, by implication, human—function, from Erasistratus, Herophilus, and Galen through Harvey and Haller, had been sporadic. Never had a continuous research tradition been established. The role of experiment had more often been the demonstration or illustration of views formed on other grounds than the language in which problems were formulated and the means by which they were resolved. From about 1820 the ideal of experimental determination of the conditions of living phenomena achieved an ascendency in physiology from which it has never been displaced through successive transformations of physiological theory. In this perspective the conceptual and theoretical changes of the science in the preceding two decades are of secondary importance, even when intimately associated with experimentalism in the work of Magendie's generation.

On 24 April 1809, Magendie read to the First Class a memoir in which he reported the results of animal experiments carried out to determine the physiological mode of action of a poison recently brought back from Java and Borneo by the botanist of the Baudin expedition, Louis Leschenault.[1] The substance, employed by the natives to tip their weapons for hunting and for war, was an extract of *upas tieuté*.[2] Magendie allowed the extract to dry on the surface of quill-sized slivers of wood, then inserted one of these into the thigh of a healthy dog. After three minutes of calm the dog showed evidence of a general malaise, immediately followed by a convulsive contraction of all the muscles of the body and a straightening of the vertebral column. This ceased at once and was followed by a few seconds of calm. Convulsive contractions were then resumed in greater strength and duration, accompanied by accelerated breathing and a more marked straightening of the vertebral column. A further half minute of calm was succeeded by even more violent muscular contractions, which finally resulted in complete tetanus. With respiration blocked by total immobilization of the thorax, symptoms of asphyxia soon appeared. While the tetanus continued, the action of the senses and brain were unimpaired; they began to weaken only when the asphyxia became very marked. A minute of complete tetanus was followed by an equal interval of abatement. A fresh attack of general tetanic contraction was accompanied by violent jerking movements which, Magendie noted, were exactly comparable to those produced by application of a galvanic current to the spinal cord of a recently killed animal. During the subsequent final period of calm, instantaneous tetanic contractions were produced by mere contact of the hand with any part of the body. A final attack of convulsions ended in death. Opening the chest and abdomen revealed that death was due to asphyxia. Repetition of the same experiment on a horse, six dogs and three rabbits gave uniform results, the violence of the contractions varying with the strength of the individual animal.

These experiments appeared to justify the conclusion that the poison was absorbed in the wound, carried by the circulatory system to all organs, and acted on the spinal cord as an "energetic excitant, of which the effects are analogous to those determined by irritating the spinal cord by a mechanical means or by the galvanic fluid."[3] Was the *upas* really absorbed? Magendie obtained evidence for this conclusion by placing a solution of the extract in contact with a variety of absorbing surfaces. Injection into the peritoneal cavity or pleura—both lined with fast-absorbing serous membranes—resulted in very rapid appearance of the symptoms of poisoning, followed by death.[4] Mucous membranes, such as those lining the small intestine, large intestine, bladder, and vagina took up solutions of the extract much more slowly, and death resulted only after convulsive attacks that were more frequent and less intense than those following absorption by serous membranes. Mixed with aliments in the stomach, the poison was taken up even more slowly than in the case of the other mucous membranes.

Was it through the intermediary of the circulation that the *upas* acted on the spinal cord? If so, then the rapidity with which the poison acted should depend on the length and difficulty of the path it had to follow through the circulatory system to arrive at the spinal cord. Magendie confirmed that this was the case by direct injections of the solution into the jugular vein, the crural artery, and the carotid. Finally, was the spinal cord really the site of action of the *upas*? Separation of brain and cord by a section made between the occipital and the first cervical vertebra during tetanic contractions did not arrest the tetanus. In those cases in which the contractions did fall off sharply after the section, it could be shown that this was due to the cessation of the circulation, and thus of transport of poison to the cord. Simultaneous administration of *upas* in the pleura and removal of the entire cord resulted in a complete absence of contraction. Administration of *upas* in the peritoneum followed by progressive destruction of the cord after onset of contraction resulted in the progressive cessation of tetanus in those parts corresponding to the destroyed portions of the cord. Injection of solution of the extract directly into the cervical region of tbe cord produced immediate tetanus in the front limbs without at first affecting the hind limbs. Gradually the tetanus became general, reaching the hind limbs at the end of the sixth minute. By the tenth minute it had ceased in the front limbs but continued in the hind limbs; soon it ceased altogether. Finally, Magendie sectioned the vertebral column and spinal cord in the lumbar region, and injected the *upas* into the cord in the pelvic area, producing contractions of the hind limbs. Eleven minutes later he observed weak contractions in the front limbs.

Magendie concluded that the extract was a powerful stimulant of the spinal cord that could cause death in a very small dose "by determining a prolonged tetanic contraction of all the muscles to which the spinal cord supplies nerves, a contraction which necessarily suspends respiration and produces asphyxia."[5] He felt that it would be a mistake, however, to view *upas tieuté* solely as a poison. In very small doses of two or three centigrams it produced marked excitation of the spinal cord without disturbing the important functions of life, and might therefore have uses in medicine treating conditions which had their seat in the spinal cord. Although *upas* was not available in commercial quantities, there was another species of the *strychnos* family, known as *nux vomica*, which supplied an extract with identical effects and was available in substantial amounts. A third species of the same family, *Ignatia amara*, also produced effects exactly parallel to those of *upas*.

Magendie was not the first to study the action of a drug or poison by carrying out experiments on living animals.[6] By the second half of the eighteenth century physicians and savants had advanced three kinds of explanations: a drug could convey its effect through absorption into the circulation followed by action on the blood or on an organ to which the substance was carried by the blood; through direct action on the nervous system via nerves at the site of administration; or by "sympathy" (impressions passing along the nerves from the site of administration to a separate site of action).

An important factor in the acceptance of the nerve explanation was the research of William and John Hunter on the structure and function of the lymphatic system. In his *Medical Commentaries* of 1762 William Hunter gave an account of experiments carried out by his brother three years earlier that were designed to determine whether the lymphatic vessels were the exclusive agents of absorption or whether the veins might also play a role in the process. All the experiments described were carried out on portions of mammalian intestine and associated lymphatic and blood vessels. Hunter opened the abdomen of a dog and isolated two six-inch sections of the intestine by ligatures tied at each end. He poured milk into each section through a small opening in one end, and applied a third ligature to close the opening. He closely observed the veins and lacteals arising from each section. Hunter found that one lacteal which had previously been white remained white, a second which had previously been transparent became white, while all the veins retained a uniform deep color with no evidence of lightening. He carried out a number of experiments along similar lines (with variation in the technique, the solution injected, and the animal subject), and all indicated that there was no evidence that any fluid was taken up by the veins, or that any fluid passed from the arteries or veins into the intestine. John Hunter does not appear to have drawn any categorical conclusions from these investigations, but William referred to them as evidence that the veins did not absorb.[7]

With direct venous absorption apparently precluded by Hunter's experiments, the slower, indirect route of the lymphatic system was left as the only conceivable path for absorption of drugs and poisons. The inferences to be drawn in favor of nerve action explanations are represented in the works of William Cullen, widely influential in medical circles in the later eighteenth century.

Nevertheless, acceptance of the view that fast-acting drugs and poisons produced their effects directly through the nervous system was not unanimous. By the end of the century one of the bases of the nerve action theory, the view that the lymphatic vessels were the exclusive agents of absorption in the living organism, was being placed in doubt by Bichat in the *Anatomie générale.* Discussing the overall course of the lymphatic vessels, Bichat noted the apparent discrepancy between the structure of the vessels and the functions they were supposed to perform:

> However little we examine the quantity of absorbants spread over all the parts, it will be easy to conceive how enormous is the disproportion of their capacity with that of the two trunks. How is it that all the serosity contained on the serous surfaces and in the cellular tissue, all the residue of nutrition, all the fat, medullary juice and synovia, all the drinks, all the product of solid aliments which ceaselessly enter into the circulatory torrent have to pass, in order to penetrate it, across such small vessels? This observation has struck all the authors: admittedly it offers a great difficulty to resolve.[8]

A number of arguments had been advanced in favor of the view that the veins also absorb, Bichat noted, but several of these were of doubtful validity. It had been observed that animals had survived obliteration of the thoracic canal, but care had not been taken to determine whether the great right lymphatic was also obliterated. Bichat pointed out that the question could be decided by tying the thoracic canal during digestion, an experiment that would "throw great light on the general question of absorptions."[9] Bichat concluded that the relative roles of the lymphatic and venous systems in absorption remained obscure and uncertain, and called for new investigations to resolve the problem.[10]

Bichat's challenge was taken up by Magendie in a "Mémoire sur les organes de l'absorption chez les mammifères" read at the Institute on 7 August 1809, less than four months after the first memoir on drug action.[11] Magendie presented the researches of this memoir as a direct development of those on the physiological action of the *strychnos* extracts. Impressed by the extremely rapid action of these substances, he had been led to question the generally accepted view that any absorbed substance must first traverse the vessels and glands of the lymphatic system before entering the bloodstream. Magendie also questioned whether poisonous substances could pass through the "tortuous and difficult" route of the lymphatic glands without undergoing any alteration. Without citing Bichat, Magendie alluded to the eighteenth-century controversy over whether the lymphatics were the exclusive agents of absorption or shared this faculty with the veins. He pointed out that an earlier opinion, shared by such figures as Boerhaave, Haller, Meckel, Frederik Ruysch, and Jan Swammerdam, had favored a role for venous absorption, a role supported by "various circumstances of structure and several physiological and pathological facts." It was only after the anatomical elucidation of the lymphatic system and the experiments of the Hunters and others that the lymphatics came to be viewed as the exclusive agents of absorption.

More recently, "a series of interesting experiments, undertaken and carried out a few years ago at the veterinary school of Alfort" had lent new probability to the notion of venous absorption. Magendie was referring to the work of Pierre Flandrin, veterinarian, anatomist, and professor at Alfort until his death in 1796. Flandrin had published his research on venous absorption in a series of four articles that appeared in 1790–1792.[12] Flandrin was unable to repeat Hunter's experiments on the uptake of colored substances such as indigo by the intestinal lymphatics. On the contrary, he found evidence that indigo included in the diet is taken directly into the circulation, but not into the lymphatics. In other experiments, he tied or otherwise blocked the thoracic canal (the vessel emptying fluid from the lymphatics into the circulatory system), and compared the results to those observed in cases of varicose lymphatic vessels seen in veterinary practice. Horses were not adversely affected by either condition. Using animals condemned because of exposure to glanders, Flandrin sacrificed each animal during digestion. He found that blood from veins associated with a given portion of intestine had the qualities of taste and odor of the digested aliment in that portion. Fluid in the associated lymphatics was clear, odorless, and tasteless. Finally, Flandrin showed that fluids could be introduced into the intestinal canal by injection of associated veins and arteries but not by injection of associated lymphatic vessels. He concluded cautiously that if his experiments did not suffice to prove venous absorption from the alimentary canal, they did throw doubt on the notion of lymphatic absorption from the canal.

Flandrin's results, apparently unknown to Bichat, opened Magendie's mind to the possibility of venous absorption, and probably predisposed him to this view. Subsequent research had, however, undermined the force of one of Flandrin's points: that tying the thoracic canal did not necessarily result in death. Aiming to clear up the ambiguity of earlier results, Dupuytren had recently carried out experiments in which he tied the thoracic canals of horses. He found that after five or six days some of the subjects had died, while others showed no ill effects. On opening the cadavers of both groups, he found that with those that had died it was impossible to make any fluid pass by injection from the inferior part of the canal into the subclavian vein. It was easy to do so with those that had survived, due to the existence of secondary lymphatic vessels which bypassed the thoracic canal and communicated directly with the subclavian vein. The mixed results of tying the canal were thus easily accounted for: in the first case the flow of chyle into the circulation had been stopped; in the second it had continued by paths other than the thoracic canal. Magendie himself had examined the secondary communications on one of Dupuytren's subjects. Dupuytren's results tended to reconfirm the important role of the lymphatics in intestinal absorption.[13]

It was with this suggestive but inconclusive background that Magendie introduced his own experiments, undertaken with Delille, on lymphatic and venous absorption. He began by asking whether ligature of the thoracic canal would prevent passage of the *strychnos* poisons to the circulatory system, and thus block its effects on the spinal cord. Tying the thoracic canal of a dog, he placed a solution of *upas* in the peritoneum. The usual effects followed with the same speed and intensity as in subjects in which the canal was not tied. He obtained similar results with other subjects when after tying the canal he placed the poison in the pleura, the stomach, the muscles of the thigh, and other locales. In no case did tying of the canal prevent action of the poison. No certain conclusions could be drawn from these results, however, since Dupuytren had shown that the thoracic canal was not always the sole point of communication between the lymphatic system and the venous system.

The aim of the second set of experiments was to isolate an absorptive surface from any connection with the circulatory system via the lymphatics. A portion of a dog's intestine was withdrawn from the abdominal cavity through an incision, and two ligatures applied to this section at a distance of about twelve inches (*quatre décimètres*) apart. Since the dog had eaten a full meal seven hours earlier, the lymphatic vessels arising from the section of intestine were full, white, and easily identifiable. On each of the lymphatic vessels arising from the section of intestine two ligatures were applied one centimeter apart, and the vessels cut between the ligatures. Great care was taken to destroy in this way all communication between the section of intestine and the rest of the body via the lymphatics. Four of the five mesenteric arteries and veins supplying the separated portion of intestine were similarly tied and cut, leaving a single artery and vein. The two ends of the portion were then cut, isolating it entirely from the rest of the small intestine. The single remaining artery and vein were then isolated from surrounding tissues along a length of four finger-breadths, and the associated cellular tissue was removed in order to be certain that no hidden lymphatics remained. A small quantity of *upas* was then injected into the isolated fold of intestine, appropriate precautions were taken to prevent leakage, and the portion, covered with fine linen, was replaced in the abdominal cavity. After six minutes the usual effects of the poison appeared with the usual intensity. Postmortem examination revealed that all ligatures remained intact, and that no poison had escaped into the abdominal cavity. Magendie concluded from this experiment, repeated several times with the same results, that the lacteal vessels were not the exclusive organs of intestinal absorption.

Reasoning that this nonlymphatic absorption might be peculiar to the intestines, Magendie designed an experiment to test its occurrence elsewhere in the body. On a dog previously anesthetized with opium, he performed an operation in which one thigh was separated from the rest of the body, leaving intact only the single crural artery and vein. As in the preceding experiments, Magendie isolated a four-centimeter length of each vessel and removed the surrounding cellular tissue. When he placed a small quantity of *upas* in the paw, symptoms of poisoning appeared within four minutes and death occurred within ten minutes. To remove any suspicion that the poison might reach the circulation via small lymphatics hidden in the walls of the blood vessels, Magendie repeated the same experiment under slightly modified circumstances. Having isolated the thigh from the body save for the crural artery and vein, he inserted a small quill tube into the crural artery. He tied the associated portion of artery to the quill by ligatures at both ends, and removed the section between the ligatures by circular incisions. He performed the same operation on the crural vein, leaving the blood transported by these vessels as the sole remaining communication between thigh and body. Placement of poison in the paw was followed by the usual symptoms in about four minutes. Joining this result to those of the preceding experiments, Magendie felt justified in concluding that the lymphatic system was not, at least in certain cases, the exclusive route taken by foreign substances in arriving at the venous system, and that this circumstance explained the rapidity of absorption and action of certain substances. This categorical conclusion was slightly qualified by the admission that the experiments reported did not resolve the question of whether it was the small endings of the veins or the lymphatic capillaries which (so Magendie assumed) anastomose directly with them that first drew up the poisons at the site of administration.

Magendie went on to point out that the blood of an animal manifesting signs of poisoning by *upas* could be considered to contain a certain quantity of the poison—to be, in fact, poisoned blood. Would this poisoned blood produce similar effects when introduced into the circulatory system of a second, healthy animal? To answer this question, Magendie arranged a transfusion of the arterial blood of an already poisoned animal to the jugular vein of a healthy animal. The transfusion lasted more than twenty minutes, during which time the blood of the donor animal changed from bright red to dark as the *upas* produced asphyxia. Repeated several times this experiment always produced the same result: no trace of poisoning in the recipient animal, which manifested only the symptoms associated with "the ordinary transfusions," namely an accelerated respiration and abundant pulmonary exhalation. To eliminate the possibility that the passage of the poisoned blood of the donor animal through the lungs destroyed its poisonous quality, Magendie carried out a second set of experiments in which he transfused blood from the jugular vein of the poisoned animal into the venous system of the recipient. He began the transfusion before the appearance of symptoms of poisoning and continued it until the death of the donor animal. He observed no effects of *strychnos* poisoning in the recipient animals, though all received a large quantity of poisoned blood. From the whole set of transfusion experiments Magendie drew the categorical (and mistaken) conclusion that the blood of animals poisoned by the *strychnos* extracts was incapable of producing harmful effects in other animals.

A number of fundamental characteristics of Magendie's physiology are already evident in his 1809 memoirs on drug action and absorption. Most striking is the surgical character of his attack on physiological problems. This is reflected in an active intervention in and manipulation of bodily processes, extensive utilization of operative skills, close attention to gross anatomical relationships, and a refusal to make categorical assertions regarding phenomena not accessible to the senses. Closely associated with this phenomenalism is Magendie's severity in assessing experimental results and his caution in formulating final conclusions. The physiology of these two memoirs is also strongly analytical. Though each of the major topics under investigation—the mode of action of powerful poisons and the relative roles of lymphatic and venous absorption—had its own background in eighteenth-century research, in neither case was the initial choice of problem dictated by general doctrinal or systematic interests. Each investigation focuses on a discrete, strictly delimited problem. These problems are self-contained in the sense that they are investigated with little or no reference to other physiological problems or to the organism as a whole, while the results of each investigation are held to stand or fall by themselves, without reference to their bearing on some more general physiological doctrine.

It is a physiology that bears the stamp of Magendie's education in the reformed medical curriculum of the First Empire, and especially of his training in surgery and anatomy. It is a physiology not different in character from that of Part 2 of Bichat's *Recherches physiologiques*, from which Magendie must have drawn inspiration despite the absence of any direct reference to Bichat. The use of Bichat as model is most evident in the transfusion experiments of the absorption memoir, in which not only the technique but also the idea of producing poisoning in a recipient animal by means of the blood of a donor exactly parallels the production of secondary asphyxia in the *Recherches*. The technique of the experimental study of absorption through isolation of a portion of intestine was borrowed from another surgeon-physiologist, John Hunter. A not insignificant role was played in the shaping of the physiology of the second of these memoirs by the veterinary school of Alfort, both through the experimental investigations of one of its faculty members, Pierre Flandrin, and through the provision of a locale and subjects for the animal experiments of Magendie and Dupuytren.

The First Class of the Institute, where Magendie originally read his memoirs, gave them a mixed reception. The report on the drug action experiments, submitted a month after the reading and signed by Raphael Sabatier, Pinel, Jussieu, Charles Mirbel, and Jacques Labillardière, was entirely favorable and, but for the dangerous nature of the contents, they would have recommended publication in the *Mémoires des savans etrangers*.[14] The report on the absorption memoir, signed by Cuvier, Philippe Pelletan, Portal, and Pinel, was delayed for almost four years, not appearing until January 1813. Magendie was praised for widely varying his procedures, considering the results of his experiments from various points of view, and suspending judgment in obscure or equivocal cases. Nevertheless, the commissioners found that his conclusion was still premature and that his facts had not yet been sufficiently multiplied or adapted directly enough to the proposed end to conclude that the lymphatic vessels were not always the route followed by foreign matters to arrive at the blood system. In particular they pointed out (as Magendie recognized) that his experiments did not resolve the question of whether it was the fine terminations of the veins or the lymphatics that first absorbed the poison, since these terminations were in "continuous interlacement" with one another. The commissioners finally recommended abstention from any definitive judgment pending further experimental inquiries promised by Magendie, and printing of the memoir with the *Savans etrangers*.[15]

Magendie's next published discussion of the absorption problem appeared in the first edition of his *Précis élémentaire de physiologie* (Paris, 1816–1817). He was unable to add any substantial evidence in support of venous absorption to that supplied by the experiments of the 1809 paper. He was able to strengthen his case by default through a more extended critique of arguments used in support of a general lymphatic absorption. Nowhere did Magendie address the difficulty underlined by the report of the Institute— namely, the possible continuous interlacement of the fine terminations of veins and lymphatic vessels. Nevertheless—perhaps because he was writing an elementary textbook—his conclusions were more categorical than the ones he had presented in his 1809 paper. He now maintained that it was certain the chyliferous vessels absorbed chyle; that it was doubtful they absorbed anything else; that it was not demonstrated the lymphatic vessels had the absorbent faculty; and that it was proved the veins enjoyed this property.[16]

The *Précis* also reintroduced Magendie's critique of Bichat's vital-properties doctrine, now in the particular context of the absorption problem. In the 1809 memoir on absorption, there had been no mention of the mode of action of the walls of venous or lymphatic vessels in allowing the passage of fluids, or of the bearing of vital-properties ideas on this question. In the *Précis* he noted the obscurity surrounding the mechanism of absorption on this level, and remarked that the attribution of a peculiar form of sensibility and contractility to the absorbent openings merely provided a thin veneer to ignorance.[17] He recommended that for the moment judgment should be suspended on the cause of the absorption of substances by the veins, but added that "it is also quite probable that this introduction is effected in another manner" than that envisioned by vital-properties doctrine.[18]

In making the latter remark, Magendie did not specify the approach he intended to adopt. Nevertheless, the direction his interest in the circulatory system was then taking is clearly indicated in other remarks not bearing directly on the absorption problem. Discussing the passage of blood from the smaller to the larger veins, he pointed out that the overall volume of the former is greater than that of the latter. Therefore, the physiologist could invoke the hydrodynamic principle that in a closed system of full tubes speed of flow diminishes with increase in size of the tube, and increases with a decrease in size.[19] A few pages later, he noted that the friction of the blood against the walls of the vessels and its own viscosity must modify the flow of blood in the veins, but that in the present state of physiology and hydrodynamics these effects could not be adequately characterized.[20]

Magendie was more assured about another aspect of his developing view of the circulation as a physical system: that the constriction of arterial and venous walls associated with the passage of reduced quantities of blood was due to their purely physical property of elasticity, and not to any muscular contraction of the walls.[21] Such elasticity solved the complex hydraulic problem presented by the distribution of fluid at variable quantities and speeds in a single system of tubes. "The elasticity of the arterial walls," he pointed out, "represents that of a reservoir of air in certain pumps of alternating action, which nevertheless furnish the liquid in a continuous manner; and in general it is known, in mechanics, that every intermittent movement may be transformed into a continuous movement by employing the force that produces it to compress a spring which reacts in turn with continuity."[22] He was clearly moving in the direction of giving as strongly physical an interpretation as possible to the phenomena of the circulatory system. It would be only a matter of time before he subjected the problem of absorption across lymphatic and venous walls to the same approach.

Magendie's absorption studies were broadened in another direction with the publication in 1819 of a memoir on the lymphatic vessels of birds. Here he reported that three years earlier he had happened to read a letter from G. Hewson to John Hunter, printed in the *Philosophical Transactions* for 1768, in which Hewson had announced his discovery of lymphatic vessels in the intestines as well as the neck of birds. It struck Magendie as surprising that an anatomist as skillful as John Hunter had seen the lymphatics of the neck but missed those of the intestines. Undoubtedly, the significance of Hunter's "failure" was increased by Magendie's prior suspicion that it was no failure at all, although he did not say this. In any case, Magendie's own dissections of more than fifty birds of all genera had convinced him that the chyliferous vessels and the thoracic canals did not exist in birds and that the only traces of lymphatic vessels were to be found in the neck of the goose. If so, absorption could be effected only by the veins.[23]

Here is the real interest of Hunter's "failure." Magendie saw in it the opportunity to provide further evidence from comparative anatomy for his views on venous absorption. His argument was truly comparative: it asserted that because a function was carried only by one set of organs at a lower taxonomic level, this same set of organs was sufficient to carry out the functions at a higher taxonomic level. The inference was not logically necessary, but depended on a kind of sufficient reason in comparative anatomy and physiology.

In the same memoir, Magendie referred to Cuvier as not having undertaken investigations on the lymphatics. In another memoir of the same year, in which Magendie revealed the existence of some new structures in birds and reptiles uncovered in his study of the lymphatics, Cuvier was mentioned again. This time the connection was direct and personal, since Cuvier had supplied Magendie with anatomical material. It may be that Cuvier had played a role in directing Magendie's investigations on lymphatics into comparative channels. Perhaps because of the medical origins and direction of his interests, Magendie ordinarily restricted his physiological studies to man, and to the mammals insofar as they paralleled man in function and structure. That his comparative anatomical studies were a departure from his ordinary course tends to strengthen the probability that he was encouraged in them by Cuvier.

Ultimately, Magendie's own results proved to be faulty, and Hewson's to be justified. In a later report to the Academy of Science on a memoir of the Strasbourg anatomist Thomas Lauth, Duméril and Cuvier supported Lauth's claim to have demonstrated and described a lymphatic system in birds, including chyliferous vessels. Magendie, by this time a member of the Academy and named as one of the commissioners to examine Lauth's memoir, continued to question Lauth's demonstration of chyliferous vessels in birds and did not sign the report. In appears that in this case Magendie, the self-proclaimed empiricist, was the victim of his own preconceptions. Already convinced of venous absorption on other grounds, he allowed his search for supporting evidence to affect his perceptoin of the anatomical reality.[24] That this could be so reveals the extent to which Magendie's research could take on a committed and directed character, a quality not allowed for by his formal methodological position.

Magendie was able to move with more assurance and success in demonstrating the physical character of venous and arterial absorption. The first results of the investigations promised in the *Précis* were read to the Academy in October 1820.[25] Generally received opinion supposed the existence of imperceptible absorbent roots, orifices, or mouths, endowed with the property of drawing up or absorbing substances with which they were in contact. Moreover, Magendie continued, in the heavily ironic tone he sometimes deployed against views that appeared to him unfounded, "they [the orifices] do not do this without discernment: on the contrary, they have an extremely fine touch; they choose with severity what must be taken or rejected, and it is only after making a due inspection that they decide to exercise their *absorbent power.*"[26] However pleasing such images might be to the imagination, they could not be a substitute for positive knowledge of the mechanism of absorption.

What would be the effect, Magendie asked, of an artificial plethora of fluid in the circulatory system on the phenomena of absorption? Magendie introduced one liter of water at 40° Centigrade into the bloodstream of a dog of medium stature. He then placed a light dose of a well-known but unspecified fast-acting drug in the pleura. He observed the usual effects, but only after a delay that was several minutes longer than usual. With several other subjects the same initial conditions were followed by the appearance of symptoms in the usual time, but the effects were noticeably weaker and more prolonged. When Magendie raised the quantity of water added to the circulation to approximately two liters, the effects of the drug, which ordinarily appeared in two minutes, were not yet evident after half an hour. When he made a substantial bleeding from the jugular vein the effects appeared in a strength proportional to the amount of blood removed, indicating that the initial lack of symptoms was due to a prevention of absorption by distension of the blood vessels. An initial bleeding produced a marked shortening of the time of onset of symptoms. To remove the objection that it may have been a change in the nature of the blood when mixed with water rather than simple distension that had prevented absorption, Magendie made a substantial bleeding and replaced the blood lost by an equal quantity of water at 40° Centigrade. The effects of the poison followed in the usual time, as if no substitution had been made. In words that might have come from the better-known pen of his student Claude Bernard, Magendie reflected that "henceforth I became, so to speak, master of a phenomenon which until then had been an impenetrable mystery for me. Able to oppose its development, produce it, make it fast, slow, intense, weak, its nature could not easily escape my investigation."[27]

Here what Bernard later characterized as the ultimate mission of experimental physiology—the operative control of the phenomena of the living organism—became momentarily explicit in Magendie's account of his own investigations. It is one of the deepest impulses of Magendie's physiology, linking it to Bichat's experimental physiology and to the clinical and operative surgery associated with it, and separating it from any merely discursive or descriptive physiology, such as that embodied in Bichat's doctrine of vital properties. Such constancy and regularity in the phenomena pointed to a physical rather than a vital action. And "among the conjectures that might be permitted in this regard," Magendie continued, that which made absorption depend on a capillary attraction exercised by the vascular walls was undoubtedly the most probable; in any case, it appeared to be consistent with all the known facts.

Stronger evidence was required, however, and this Magendie offered in the form of further experiments. He drew the general conclusion that the walls of all the blood vessels, arterial and venous, dead and living, large and small, possessed a physical property that would account for all the phenomena of absorption. He did not yet go so far as to claim that this was the only property they possessed, but did maintain that it was consistent with all known facts. Magendie further specified the nature of the physical process involved as a capillary attraction, but did not offer special evidence for this view. He expected that his own results would find numerous applications in pathology and therapeutics.

In a favorable report to the Academy on Magendie's memoir, the commissioners Thenard, Gay-Lussac, and Berthollet noted that they had witnessed his experiments, but they refrained from passing any explicit or final judgment on Magendie's further conclusions on the physical character of absorption, and specifically on the capillary attraction of the vascular walls. They did remark that Magendie had here given fresh proof of "the sagacity with which he attempts to bring into physiology the rigorous method of the physical sciences, and to banish that which is hypothetical."[28]

As we have seen, this was not the first—nor would it be the last—instance in which Magendie offered his work to the Academy and the Academy passed its judgment, usually very favorable, on that work. The cumulative impression gained from a reading of the *Procès verbaux* over the period 1809–1820 is of a strong interaction between the aspiring physician-savant and the elite of French science. From his earliest published papers, it was for the Academy, not for the Paris medical community, that Magendie wrote. The project of "Quelques idées sur les phénomènes particuliers aux corps vivans" was to make physiology into a science worthy of a place beside the exact sciences of astronomy, physics, and chemistry. After these sciences, the recent successes of comparative anatomy fired Magendie's imagination.

The Academy, for its part, played the dual role of confirming or questioning Magendie's results through logical criticism or repetition of his principal experiments on the one hand, and reinforcing his tendencies or suggesting new directions of investigation on the other. The Academy's commissioners occasionally questioned the adequacy of Magendie's evidence, or held themselves aloof from commitment to his conclusions. In general, however, they were impressed with his boldness in questioning generally accepted views, his caution and rigor in formulating and supporting his own, and his avoidance of excursions into hypothetical explanation. The anatomical precision he brought to physiology was especially appreciated. Commenting on Magendie's masterful analysis of the nerve supply of the muscles of the larynx in his memoir on the use of the epiglottis in swallowing, the commissioners Pinel and Percy remarked that "exactitude in anatomy is the surest path to truth in physiology."[29]

The same report opened with several enthusiastic paragraphs on the current quality and direction of physiology in France, in which praise of Magendie was mixed with a measure of self-congratulation:

> We cannot witness without interest an enlightened, judicious physiologist, a friend of truth, bring to bear a methodical doubt on points of doctrine consecrated by common belief, practice on them a sort of censure and revision, submit them to rigorous experiments, and seek to fix opinions which for too long have been wandering or undecided. The time has passed when the teaching of physiology was composed of hypothetical explanations and in which books were filled with purely imaginary systems . . .
>
> How different is the physiology of our day from that of the last century! By what precious and unexpected discoveries has it been enriched in the last twenty years! And it is above all to the French that this surprising progress is due . . .
>
> It must be acknowledged. Among us the exact sciences, cultivated with so much glory and success, have given a better direction to thought. The need for and severity of analysis have changed ideas and perfected judgment, and it is well known which institution has most claim to glory for having produced this great movement and given this useful example.

No one has been more faithful to its principles than M. Magendie. He has presented himself to us only surrounded by demonstrations, and all the memoirs which he has offered to the Class consist of proofs and facts.[30]

For all their underrating, not to say caricature, of earlier physiology, Pinel and Percy here give expression to a fundamentally accurate perception of the new direction recently taken by physiology in France, of the role played by the Institute in this change, and of Magendie as its most representative figure. Their remarks represent one of the earliest evidences of a rising consciousness of the rapid scientific development of physiology. Though they do not appear to be aware of the fact, they were also contributing to that development by consistent use of the terms "physiology" and "physiologist," for it was in the Institute in the period 1800–1820 that these terms first became standardized and gained official recognition in Europe.

Active interest in and support for Magendie's physiological research came from the highest levels of the Academy. Cuvier was among the commissioners assigned to report on Magendie's earliest papers on drug action, absorption, and vomiting, and it was undoubtedly on his suggestion that Magendie was encouraged to study the phenomena of vomiting in birds and other animals lacking a diaphragm, and in general to expand his physiological studies in a comparative direction.[31] When, early in 1814, Magendie was called to military service for the third time, he owed his exemption to a special decree obtained by the intervention of the First Class of the Institute, acting on his behalf.[32] Magendie's physiology had drawn the attention of Laplace and Berthollet, and it was partly on their recommendation that he obtained a position with the *Bureau central des hospices* of Paris in 1818.[33] It was also through Laplace as intermediary that the Academy received, in 1818, from an anonymous donor (later identified as Antoine Auget, baron de Montyon) the proposition to establish a prize in experimental physiology. It is likely that Laplace himself was behind the choice of field. Like statistics, for which the baron de Montyon had already founded a prize, experimental physiology was now viewed by the Academy as a developing area of great potential.[34] No doubt Laplace and Berthollet were attracted to Magendie's project to shape a physiology on the model of the physical sciences, and their encouragement and approval were an important factor in the expansion of his interest in physical and chemical approaches to physiological problems between 1809 and 1820.

In electing Magendie to membership in 1821, the Academy was at the same time endorsing the kind of physiology he represented and assuring it a place of influence in the world of French science. The acquisition of a foothold within the Academy represented a seemingly small but in fact highly significant movement away from the almost exclusively medical institutional context in which physiology had been cultivated three decades earlier.

However important such a change may have been, it represented only a first step in the institutional differentiation of physiology as a science in France. The stresses involved in the introduction of experimental physiology into the Academy in the 1820s and the fragility of its independent status in that institution are reflected in the records of the *Procès verbaux*. When the First Class of the Institute was organized in 1795, no one of its ten sections was explicitly designated for physiology. Aspects of what would later be termed human or animal physiology had to find a place in several of the sections established at that time, particularly Chemistry (V), Anatomy and Zoology (VIII), Medicine and Surgery (IX), and Rural Economy and the Veterinary Art (X).[35] In 1800, shortly before the publication of the *Recherches physiologiques*, Bichat was listed as an anatomist among the candidates for a place in the section of Anatomy and Zoology.[36] In 1811 the student of Bichat and experimental physiologist Pierre Nysten was listed as a physician among the candidates for a place in the section of Medicine and Surgery.[37]

It was not until 1821 that the issue of the representation of physiology within tbe existing structure of the First Class became a question for conscious reflection and open debate. In July of that year a place became vacant in the section of Anatomy and Zoology. Two members of that section, Geoffroy Saint-Hilaire and Pierre Latreille, presented a list of zoologists as candidates for the position. A counter list was offered by two other members of the section, Pinel and Duméril, who wished to "preserve the right to present a list of persons occupied with human anatomy." Accordingly their list was in two parts, with one column for zoologists and one for anatomists. Among the latter were Magendie and

Edwards, both of whom were primarily involved in research in experimental physiology. An extended debate ensued in the Academy between the adherents of Geoffroy, who read a memoir in support of the view taken by himself and Latreille, and those of Duméril, who responded.[38]

An extract of Geoffroy's memoir was printed the following year in the *Revue encyclopédique*.[39] Geoffroy's basic intention was to preserve the section of Anatomy and Zoology for zoologists, while according full recognition to the recent progress of physiology and its need for fuller representation within the First Class. Zoology, he felt, was already underrepresented in the First Class, and it would be a disaster if medical doctors and human anatomists were now to be admitted to its section on an equal basis, thus reducing the role of zoologists still further. Moreover, he pointed out that the financial and honorific resources available to doctors outside the Academy were far greater than those to which zoologists had access. At the same time he conceded that, on the basis of its title and prior practices, the section of Medicine and Surgery was equally ill suited to receive human anatomists and physiologists. In fact, Geoffroy continued, the problem had to be traced to the original organization of the First Class in 1795: "Physiology was then being born. Although Haller had already thrown on it all the light of his genius, it was difficult for this science to take root in the corporations. Still not introduced into the former Academy of Science, it was omitted in the formation of the Institute: it is this oversight that they wanted to repair in some way by directing the generosity of an anonymous person to found an annual prize in experimental physiology."[40] Since that time, the seriousness of the original omission had been magnified by the continuing progress of human anatomy and physiology.

> The embarrassment in which we find ourselves is therefore due not only to the original imperfection of our statutes. It goes higher; it results from the perfection of the sciences, and above all from the impulse of the minds which carry them, with an activity formerly unknown, toward the great science or knowledge of the laws of life.
>
> This embarrassment results from the fact that, in the formation of the Academy, these brilliant acquisitions of the human mind were not foreseen. There is no doubt that you owe a reception— and a very great reception—to these new sciences.[41]

Only let that reception not be at the expense of zoology. Such was Geoffroy's message to the Academy.

In a final footnote, Geoffroy went on to report that several of his colleagues had indicated a willingness to form a separate section of physiology, provided there would be no increase in the total number of members of the Academy.[42] By subtracting one member each from the six members allotted to five of the existing sections, a new section of five members could be created for physiology. According to this revised classification, the sections of Mineralogy, Botany, Rural Economy, Zoology and Anatomy, Medicine and Surgery, and Physiology would each have five members, keeping the total membership at thirty. Such a reorganization would provide ample room for the new field, while minimizing the cost to the older sections.

The scheme proposed by Geoffroy and a few others was not adopted, however, and the existing structure of the sections was preserved. A zoologist was chosen for the place in the section of Anatomy and Zoology, and Magendie had to await another opportunity. It came in November 1821, when a place became vacant in the section of Medicine and Surgery. Candidates were divided into two categories: "physicians," including Chaussier and René Desgenettes, and "authors of works useful to medical science," comprising Magendie and Matthieu Orfila. Magendie was elected on the second vote.[43]

In lieu of a reorganization of the First Class creating a separate section for physiology, Magendie had to attempt to make a place for his field within the section of Medicine and Surgery. When in 1828 a place became available in that section, its members, acting through Magendie, presented a double list of candidates designated as "physiologist-physicians" or as "practicing physicians."[44] That such ad hoc compromises and circumlocutions were still necessary seven years after Geoffroy's address to the Academy is a telling indication of the distance that remained to be covered if physiology was to gain a secure institutional base in the world of French science.

The chair of physiology (until 1822 physiology and anatomy) at the Faculty of Medicine was a teaching rather than a research post, and its principal holders between 1794 and 1858—François Chaussier, André Duméril, and Pierre Bérard—made no important contributions to experimental physiology.[45] The first unambiguous step in the direction of the institutionalization of experimental physiology in France came only in 1854, with the creation of a chair of general physiology for Claude Bernard at the Sorbonne. In the meantime, the practitioners of the field had to find places for themselves within preexisting institutional frameworks such as the chair of medicine at the Collège de France (Magendie, Bernard) or the chair of comparative anatomy at the Muséum (Flourens). As long as this was the case the position of the field was highly precarious, since its institutional continuity was not assured beyond the duration of the interests and lifetimes of the individuals who happened to cultivate it. Even after 1854 Bernard was long engaged in running skirmishes with zoologists over the independent existence, definition, and territorial claims of his field.[46]

In timing and content, the ascendency of experimentalism signaled by Magendie's entrance into the Academy owed much to the example of Bichat. The impression is strengthened by a contemporary account of Paris medicine left by the English traveler and physician John Cross. Writing of his visit to the medical institutions of Paris and Montpellier in the years 1814–1817, Cross noted with some exaggeration that Bichat's works had become for the physiologists of the capital "a species of holy scripture from which one cannot depart without sacrilege." Bichat's example had been especially influential in producing a "mania for vivisections," a fondness for isolated observations, and an aversion to general theories. Cross, who was inclined to favor the views of Barthez, obviously disapproved of the trend.[47]

What Cross overlooked, and what is evident from an examination of the careers of the members of Magendie's generation, is the role of factors peculiar to the France of the Revolution and First Empire: the new medical education; the mood of expansive optimism and ambition among the younger members of the scientific and medical community; the continuing professionalization of the healing arts and their consequent association with the sciences; and the presence of the First Class of the Institute and its openness to the medical sciences. All helped foster a generation whose members practiced a physiology sharing essential features with that of Magendie. His name survives because, of that generation, he was most articulate in defining the field and its goals, most persevering and prolific in prosecuting its researches, and most successful in achieving recognition from the French scientific community.

Although recent scholarship has affirmed the pivotal importance of the first two decades of the nineteenth century in France for the origins of modern physiology, it has denied a central place to experimentalism. William Albury sees Magendie, as he does Bichat, in the context of a putative transmutation of European thought in the decades around 1800.[48] In his view, Bichat stands clearly on that side of the watershed that remains immersed in the essentially taxonomic thinking of the eighteenth century. Magendie, in contrast, is the first physiologist to embody in his science the new order of biology and its commitments to study of the whole organism in terms of the integrated and adapted structure and function of all its parts. As protégé of Cuvier, Magendie imports into physiology Cuvier's holistic conception of the organism. For Albury, such a conceptual shift entails a new concept of function freed from the bonds of a one-to-one association with individual structures, whether tissues or organs, and emphasizing instead the cooperation of several such structures in the accomplishment of a common end. This notion of function in turn impels recourse to experimental methods in the manner of Magendie. The result is not to deny a place of significance to experiment in the new physiology, but to subordinate experiment by making that place a consequence of conceptual changes alone.

Albury is correct in ascribing to Cuvier an influence on Magendie, particularly in regard to the latter's critique of Bichat's vital-properties doctrine. Yet, as we have seen, Magendie's concept of organization cannot be identified with Cuvier's holism, but is reminiscent of that found in Bordenave, Bichat, and other writers of the late eighteenth century. Magendie's formal definition of function may also be found in Bichat's writings, and was implicit in much of physiology from at least the time of Harvey. The deductive chain that leads from Cuvier's holism through its expression in Magendie's physiology, through the latter's concept of function to his necessary use of experiment, therefore breaks down. And Albury's concentration on conceptual and methodological commitments alone means that the essential kinship of the experimental physiologies of Bichat, Nysten, Dupuytren, Legallois, and Magendie and the basis of that kinship in a common pattern of training and institutional association are not and cannot be perceived.

A less deductive but equally clear emphasis on concepts at the expense of experimentalism as the key to the new direction of physiology has been advanced by Michael Gross.[49] Gross keys the change in physiology in the period 1800–1825 to a fundamental reformulation of the concepts of sensibility and contractility. Whereas eighteenth-century vital-properties doctrine had almost always seen those properties as diffused throughout the organism, taking different forms in particular organs and tissues, in the new physiology they are localized in the nervous and muscular systems. This localization involves a thorough critique of vital-properties doctrine, a critique especially marked in the work of Magendie, Legallois, and (later) Flourens. In Gross's view, the reformulation of the notions of sensibility and contractility by physiologists was prompted by the example of the zoologists, who had long allotted privileged roles to the nervous and muscular systems. The principal spokesman for the zoologists was Cuvier who, working through the First Class, was well placed to encourage the new direction in the thinking of his physiological colleagues. In Gross's view, these changes were not dependent on the use of experiment, and the latter was a coincidental accompaniment to the transformation of concepts.

The reformulation of concepts described by Gross is a real one, and he rightly emphasizes its prominence in physiology in the first two decades of the century. His case for the role of the zoologists, and of Cuvier in particular, in this reformulation is also persuasive. Yet the localization of sensibility and contractility was only one among several issues engaging the attention of physiologists. Legallois's studies on the way in which blood changes in the organs, Nysten's investigations on the effects of gases introduced into the circulatory system, Dupuytren's work on the role of nerves in respiration, and Magendie's research on drug action are only a few of the problems then under study that cannot be subsumed under the localization issue. What does provide a common denominator in these studies, and in those focused on the problem of localization, is the central place occupied by experiment as research method. No other theme was given such strong and repeated emphasis in the reports of the Institute on physiological research between 1800 and 1820, regardless of the special topic under scrutiny. Experiment in the pattern of Magendie and his generation was perceived by the Institute's commissioners themselves as the distinguishing mark of the new physiology.

One of Magendie's institutional associations, that with the First Class of the Institute, urged on him more extensive use of physics and chemistry as tool sciences in physiological research. By the time he wrote the first edition of the *Précis*, Magendie was moving beyond the legacy of Bichat's experimental physiology, especially in his increasing use of physical and chemical approaches to physiological problems. Whereas the expanding physical dimension of his investigations in the years before 1821 is best represented in his absorption studies, it was in his continuing research on the action of drugs on the human and animal organism that chemistry came to play the most prominent role. That this could be so was due in large part to the contemporary development of a new field of chemistry, centered in Paris and associated with the names of two graduates of the Ecole Supérieure de Pharmacie: Joseph Pelletier and Joseph Caventou.

Knowledge of the active principle clarifies the pharmaceutical preparations of medicines, makes known rational formulas, and distinguishes them from those that are empirical, absurd, and often dangerous.

—*Joseph Pelletier and Joseph Caventou* (1820)

6   Pharmacists and Chemists

On 14 December 1818 the pharmacists Pelletier and Caventou announced to the Academy of Science the isolation of the active principle of the *strychnos* family of plant poisons, a principle which they had shown to be a salifiable organic base.[1] In the same memoir they outlined general views that would serve as the basis of research on the most active principles of plant drugs and poisons. This research, most of it to be carried out over the next few years, would enrich chemistry with a new class of compounds, the alkaloids, and would supply medicine with pure active principles that could be administered for the first time in exactly measured doses. Behind its initiation and successful prosecution lay years of training and experience in pharmacy and chemistry, a prior collaboration and continuing relationship with Magendie, and the recent isolation of the first salifiable organic base, morphine.

The interaction of Magendie with Pelletier and Caventou in their work on the first alkaloids exemplifies the way in which the professionalization of the healing arts in France had created, by the early nineteenth century, an environment that could encourage the development of experimental physiology and shape its choice of problems and tools. Organic chemical analysis had become an active field in Paris from the 1780s. In part this was the result of the new impetus given chemistry by the reforms of Lavoisier and his colleagues. That is not the whole story, however, for more specific institutional reasons must be invoked to explain the direction taken by the field. The Société Royale de Médecine, as one aspect of its official responsibility for public health and for the approval of new medicines, had given special encouragement to the analysis of medicinal and nutritional substances. After the Revolution the Paris Ecole de Pharmacie and Société de Pharmacie took over the roles of the defunct Société Royale de Médecine in promoting organic analysis, and especially the search for the active principles of medicines.

The latter task was made more urgent by a crisis of confidence in the materia medica and in drug therapy among the Paris medical community, extending back at least until the 1780s but becoming especially acute after 1800. Pharmacists perceived a threat to the body of knowledge on which their claims to a hard-won and recently consolidated professional status were based. In response, they moved to rationalize that knowledge and to provide it with a firm chemical basis. At the same time some physicians, among them Magendie, initiated efforts to overcome therapeutic skepticism through clinical and experimental research. The union of the two complementary lines of work, by pharmacists and physicians, organic analysts and physiologists, came in 1817 with the first joint publication by Pelletier and Magendie. It was a collaboration that would begin the foundation of modern pharmacology.

Joseph Pelletier, the son and grandson of pharmacists, began his formal studies of pharmacy in 1806 with enrollment in the three-year-old Ecole Supérieure de Pharmacie.[2] He was awarded his diploma as pharmacist in 1810 on the unanimous vote of his examiners.[3] Soon after taking his diploma Pelletier began to publish chemical analyses of gum resins, including opoponax, sagapenum, asafetida, bdellium, myrrh, carrana gum, and frankincense—investigations for which he was awarded the *docteur ès sciences* in 1812. Thereafter he continued his analytical studies of plant and animal products and, as a student of René-Just Haüy, wrote a second thesis on the use of physical characteristics, such as electrical conductivity, in mineralogy. When in 1815 he was named assistant professor of the natural history of drugs at the Ecole de

Pharmacie, he made use of the latter training to lecture on mineralogy as well as on his designated subject.

In 1817 Pelletier announced the successful chemical isolation of the active principle of ipecacuanha root, called emetine. This investigation, carried out with Magendie, was to be the first of an extended series of analytical studies of the active principles of plant drugs and poisons. In the same year, Pelletier began an active collaboration with another pharmacist-chemist, Joseph Caventou. Their most important joint efforts were those initiated with the isolation of strychnine in 1818. In this investigation they not only isolated in pure form the active principle of a powerful poison but also identified it as a salifiable organic base and defined a general program for the discovery of similar principles. In the next few years, the two pharmacists reported the isolation of brucine and veratrine (1819), cinchonine and quinine (1820), and caffeine (1821).

Pelletier set up a factory for the manufacture of quinine sulphate, whose utility in the treatment of several types of fever was quickly recognized. In 1825 he succeeded Pierre Robiquet as full professor of natural history at the Ecole de Pharmacie, and in 1832 was named assistant director of the school. Throughout these scientific, academic, and industrial activities, Pelletier continued to maintain the pharmacy he had inherited from his father. He died in 1842, after election to the Academy of Science in 1840.

Before he joined forces with Pelletier, Caventou's career had followed a similar path.[4] The work on alkaloids established Caventou's scientific reputation, and in 1821 he was elected to the Académie de Médecine, a year after Pelletier. In 1825 he joined the teaching staff of the Ecole Supérieure de Pharmacie, and was subsequently promoted to *adjoint* professor of chemistry in 1830 and full professor of toxicology in 1834. In 1827 he shared with Pelletier the Montyon prize of the Academy of Science for their research on quinine. Like Pelletier, Caventou managed a pharmacy in addition to his teaching and research activities.

Some idea of the kind of chemistry practiced by Pelletier and Caventou can be gained through a brief examination of the first of Pelletier's early papers. The research on gum resins that he presented for the degree of *docteur ès sciences* was initiated with an "Analyse de l'opoponax," published in the *Annales de chimie* for 1811.[5] His aim was to analyze the gum resin opoponax, found in nature as a single composite substance, into its constituent com-

pounds, to identify these clearly, and to assess their relative proportions by weight. He first separated the original substance into
two main constituents, then broke down each of these into three
principal components. In the process, he noted three minor constituents and their weights. In each step, he employed a mixture
of techniques: determination of physical properties; heating and
cooling; solvents (water, alcohol, ether); wet distillation; burning;
treatment with acids, alkalis, and salts; and tests for precipitates.
He also carried out dry distillation of the original substance, but
reported it apart from the main analytical discussions, and its results did not figure in the statement of final conclusions. He made
no attempt to do elementary analysis, even of a qualitative kind.

If there was anything novel in Pelletier's first published investigation, it lay in his choice of subject and not in the kind of results
he wished to obtain or the techniques he employed. By the time
Pelletier began his formal studies in pharmacy in 1806, the aims
and methods of the chemical analysis of plant materials had undergone almost three centuries of progressive development, in which
Parisian pharmacists and chemists had played a prominent role.[6]

Thanks in part to the work of the influential French pharmacist,
chemist, and teacher Guillaume Rouelle and his students, Paris in
the second half of the eighteenth century was a leading center for
the chemical analysis of plant and animal substances. A prominent
role in the further development of this area of chemistry was
played by the physician and chemist Antoine Fourcroy.[7] When
the Société Royale de Médecine was formed in 1776, Fourcroy
was already in close contact with its members. He was admitted
to the meetings of the Société in 1778, and was elected associate
of the Société in 1780 after qualifying as a doctor.

As a medical student Fourcroy developed a special interest in
chemistry, which he studied under Jean Baptiste Bucquet. Bucquet—from 1775 professor of pharmacy and from 1776 professor
of chemistry at the faculty of medicine, an early member of the
Société Royale de Médecine, and from 1778 a member of the
Academy of Science—was primarily interested in the application
of chemistry to natural history and medicine.[8] In his fourth medical
thesis, presented in 1779, Fourcroy argued that chemical analysis
of body fluids and solids could lead to a new understanding of
some diseases and new methods of treatment.[9]

Fourcroy's interest in medical chemistry found a favorable en-

vironment within the Société Royale de Médecine. In part this
was due to the interests of early members like Bucquet, in part
to the nature of the official tasks undertaken by the Société. When
it received its royal letters-patent in August 1778, the Société was
charged with the examination and judgment of all new remedies
to be sold in France.[10] Possessors of each new remedy were henceforth required to submit a sample to the Société, together with a
list of its purported effects, its indications, and its manner of preparation. Two commissioners, named to examine each new medicine, were to make sure that the remedy was not already listed in
a current pharmacopoeia, determine its good or ill effects, and
submit a written report to the Société, recommending approval
or rejection. Approval could be positive, in which case a *brevet*
would be awarded, or it could be tacit, in which case the Société
would merely tolerate the sale of the remedy. The commissioners
were to be chosen from among those members of the Société
having the best knowledge of chemistry. Standards were rigorous.
In its plan for the reorganization of medicine submitted to the
National Assembly in 1790, the Société reported that since its
inception it had awarded only four approbations to new remedies
and had rejected more than eight hundred.[11] As a member of the
Société with strong training and interest in chemistry, Fourcroy
was often called upon to act as commissioner. Between October
1780 and March 1791 he signed no less than eighty-eight reports
on new remedies or processes.[12]

Stimulated by its dominating concern for public health and responsibility for the supervision of medicines and mineral waters,
and by the chemical interests and knowledge of its members, the
Société Royale de Médecine undertook from an early date to
improve drug therapy through the chemical and clinical study of
medicinal substances. The first volume of the Société's *Histoire et
mémoires,* for the year 1776 (published in Paris in 1779), promised
a separate section on medical chemistry, including specifically the
analysis of alimentary and medicinal substances and mineral waters.[13]
Members of the Société contributed appropriate articles. Among
the earliest was an analysis of ipecacuanha by Joseph Lassone and
Claude Cornette, who defined the aims of the program. They
pointed out that knowledge of the properties of medicines required exact study of the principles constituting composite bodies.
"It therefore appears to us that a persevering work directed ac-

cording to these views should be of the greatest utility, since it will have as a goal, in ascertaining the virtues of remedies, making their preparation more uniform." The task was urgent, for "there is still a large number of substances, employed daily, on whose uses and effects we lack very precise ideas."[14] A "persevering work" was undertaken, and subsequent volumes of the *Histoire et mémoires* included a significant number of chemical studies of drugs, by Lassone *père et fils,* Cornette, Caille, Fourcroy, and others.[15]

The Société also offered prizes designed to encourage research along the same lines. In explaining one such prize (for chemical analyses that would throw light on the nature of antiscorbutic plants drawn from the family of the crucifers), the Société noted that "the Company desires to eliminate the uncertainty that reigns in the works of materia medica on the nature of these plants, which can be done only by a persevering work on a few of these plants, with the care and precision that modern chemistry can bring to plant analysis."[16] The prize was shared by the apothecaries Guéret and Pierre Tingry, whose memoirs appeared in volume 5 of the *Histoire et mémoires.*[17] When in 1790 the Société Royale de Médecine recommended to the National Assembly that chemistry should be one of the bases of instruction in pharmacy, that the same professor should teach both chemistry and pharmacy, and that a reform of the dispensary based on "the present state of chemical and medical knowledge" should be undertaken, it had behind it more than a decade of experience in the chemical study of medicines.[18]

It was within this milieu that Fourcroy formulated his own program for the reform of materia medica, published in 1785 as *L'art de connoître et d'employer les médicamens dans les maladies qui attaquent le corps humain.* In 1782 he had begun to lecture on materia medica to private students, a course that he would continue for three years.[19] In articles printed in the *Histoire et mémoires* for the years 1782 and 1783, he had called for an alliance of chemistry and medicine and had presented the results of chemical investigations on the alterations of animal humors in disease and on the nature of muscular tissue.[20] In *L'art de connoître* he presented a comprehensive critique of the materia medica and suggested a program for its reform. For Fourcroy, the materia medica of the day suffered from three main difficulties: polypharmacy, or the concoction of medicines out of several or many ingredients that were themselves often little understood; the separation of the study of medicines from the study of diseases; and the lack of any standard organization of the knowledge of the field. The uncertainty and ineffectiveness of drug therapy engendered by polypharmacy would be eliminated only by a vast work of simplification and analysis.[21] The gap between knowledge of the general properties of medicines and knowledge of their effects on diseases, which Fourcroy believed to be both "the most uncertain and the most neglected" aspect of materia medica, would be overcome through clinical studies.[22] Finally, Fourcroy proposed an organization in six sections that he hoped would combine the most essential contemporary knowledge in natural history, chemistry, pharmacy, and medical practice in a way that would be useful to young students.[23] In the end, only the first two sections of the projected work were published, perhaps because Fourcroy found a new focus of activity in the Academy of Science, to which he was elected in May 1785. Even in incomplete form, however, *L'art de connoître* gave greater visibility to the current of constructive reform of drug therapy that had grown up within the Société Royale de Médecine. It would not be without effect.

Fourcroy became one of the earliest partisans of the antiphlogistic theories of Lavoisier, and in 1789 he joined with Lavoisier and his colleagues in founding the *Annales de chimie,* to which he became a regular contributor.[24] With the coming of the Revolution he reluctantly became involved in politics, and his scientific activities were temporarily suspended. Not long after the resumption of normal scientific life, in October 1796, Fourcroy, along with other scientists including Berthollet and Guyton de Morveau, was elected *associé libre* of the newly formed Société des Pharmaciens de Paris.[25] Three months later he gave an address to the Société in which he recalled some of the themes of the earlier projects of the Société Royale de Médecine and of his own book on materia medica. Already in his short-lived journal of 1791, *La médecine éclairée par les sciences physiques,* he had stressed the central importance of chemistry in the purification of the old materia medica and in the discovery of new remedies, and had underlined the fundamental interdependence of pharmacy and chemistry.[26] Now he returned to the latter theme, arguing forcefully that as chemistry was the necessary basis of pharmacy, so pharmacists could and should contribute to the contemporary advancement of chemistry, as they had so often contributed to chemical knowledge in the past.[27] He urged pharmacists to take upon themselves the responsibility for the chemical analysis of medicines, whose chemical properties formed an inextricable part of their daily practice.[28]

European commercial and colonial expansion and exploration continued to yield new plants and animals, and Fourcroy emphasized the opportunity, predicting to the pharmacists that these discoveries "will become so many subjects that you will hasten to treat, and which will perhaps furnish you with unexpected results."[29] In effect, Fourcroy was asking the pharmacists to resume the work of the now defunct Société Royale de Médecine on the chemical analysis of plant and animal matters, with special emphasis on medicinal substances. In Fourcroy's view, pharmacy was to provide the new institutional locale for the Société's old project to join chemistry and medicine.

Fourcroy had already supplied an example of the kind of plant analysis he had in mind in a paper on cinchona bark, published in 1791.[30] He was unable to isolate the active principle—this was left to Pelletier and Caventou's work of nearly thirty years later—but his systematic analytical procedure and his insistence on the accurate recording of such measurable details as the quantities of solvents, the durations of extractions, and the weights of products made this paper a model of proximate plant analysis for the next two decades.

Fourcroy's work on the chemical analysis of plant and animal substances was taken up and disseminated to a younger generation of chemists by his student and collaborator, Nicolas Vauquelin. Vauquelin came to Paris in the later 1770s to seek training as a pharmacist.[31] By 1784 he had become Fourcroy's laboratory and lecture assistant. In Fourcroy's laboratory Vauquelin learned the best analytical chemistry then available, and absorbed his teacher's views on the importance of the chemical analysis of plant and animal materials and on the potential role of pharmacists.[32] Pursuing an impressive academic career, Vauquelin won appointments to several of Paris's leading technical and scientific institutions. Most important for the future of organic analysis, he was named director of the Ecole Supérieure de Pharmacie when it was established in 1803. By the middle of the first decade of the nineteenth century, Vauquelin was in a powerful position to implement Fourcroy's program for the chemical analysis of plant and animal substances, not only through his own work but also through the education of a younger generation. As a pharmacist and as director of the Ecole de Pharmacie, he was especially well placed to promote that special link between pharmacists and organic analysis called for by Fourcroy in his address to the Société des Pharmaciens. In the two decades between 1800 and 1820 Vauquelin did become, through his teaching and example, the head of a large school of analytical chemists in France. Among these were two of his students at the Ecole de Pharmacie, Pelletier and Caventou.[33]

In the first decade of the nineteenth century, the chemical analysis of drugs took on a new urgency for the pharmacists as Paris medical circles were swept by a fresh wave of calls for the rationalization and purification of the materia medica.[34] Bichat led the post-Revolution attack. In the *Anatomie générale* he was severely critical of the existing materia medica, calling it "a shapeless ensemble of inexact ideas, of often puerile observations, of illusory means, of formulas that are as bizarrely conceived as they are fastidiously assembled."[35] In the course on materia medica that he began three months before his death in 1802, he contrasted the vague, uncertain, and indeterminate character of the materia medica with "the sciences of facts, such as anatomy."[36] Materia medica would make progress only by avoiding theory and polypharmacy and by giving greater attention to pathological anatomy and to the clinical and physiological study of the effects of medicines on the body.

At the center of Bichat's course was a series of clinical trials of selected medicines, carried out in the wards of the Hôtel-Dieu. Aided by students, he administered the various medicinal substances in isolation, carefully recording their clinical effects. He then gave them in combination, two or three at a time—modest numbers of constituents relative to the old polypharmacy—to assess the new properties produced by their association. He began to read extensively in the older literature, and his course notes contain many references to earlier authors such as Cullen, Desbois de Rochefort, and Boerhaave.[37]

Bichat's short-lived effort was taken up by the physician of the Salpétrière, C. J. A. Schwilgué, who attributed the slight progress made by materia medica to the former weakness of the relevant sciences, defects in the preparation of medicines, and insufficient clinical experience. Quoting Fourcroy, Pinel, and Bichat against polypharmacy, Schwilgué noted that he had used in his own clinical trials only substances taken in isolation, in as pure a state and with as little preparation as possible.[38] Echoing Bichat's stand in the *Anatomie générale,* the physician of the Hôpital Saint-Louis, Jean Alibert, took as the basis of therapeutics "the experimental doctrine of sensibility and irritability," and asserted that therapeutics should be closely linked with physiology and pathology. Alibert sharply criticized the "inexact" and "vague" nomenclature of traditional drug therapy and attempted to introduce more precise language.[39] Finally, the physician and professor of botany at Amiens, Jean Baptiste Barbier, coined the new term "pharmacology" to designate his plan for a field that would supplant the older materia medica. The new field would base itself both on knowledge of the natural history, chemistry, and pharmaceutical preparation of each medicine and on knowledge of its effects on the organs. In contrast to therapeutics, pharmacology would examine all effects of a drug on the living economy, not confining itself to the healing effect alone. In Barbier's view, the pharmacologist would have to be a chemist in the first part of his subject, a physiologist in the second. Barbier underlined the importance of plant analysis, pointing to its recent improvement in the hands of "the modern chemists" and comparing its precision with that of anatomy.[40]

The acute sense of dissatisfaction with the traditional materia medica expressed by leading Paris physicians in the first decade of the century was not lost on their colleagues in the Société de Pharmacie, who were just then in the midst of a struggle to establish pharmacy as a profession on the level of medicine and surgery.[41] In a report on a formulary recently published by the physician Jadelot, the influential member of the Société Charles Louis Cadet took note of the recent trend among physicians to question or abandon the traditional products of the pharmacist's art in favor of simple medicines composed of one or two ingredients: "Simple substances, prepared without method, appear to have won all the votes." Although he felt that this trend was being taken to extremes, Cadet conceded that pharmacy had entered a tangible decline relative to the recent progress of medicine and surgery. He saw the "analytical method" as the key to the success of the latter two, and a regular system of classification and reform of its "absurd" nomenclature as the essential prerequisites to pharmacy's recovery. "This reform is urgent if we want to revive a dying pharmacy, since it is useless to seek to palliate an evil that we all feel. Few physicians believe in our art, and if we do not make courageous efforts, in a short while there will be many makers of medicines and not a single pharmacist."[42]

The sense of unease reflected in Cadet's report is also evident in one of the subjects proposed for a prize essay by the Société de Pharmacie in 1810. Respondents were to answer the questions "What is the present state of pharmacy in France? What part does it play in the art of healing, and what improvements is it capable of?" The author of one unsigned response felt that although pharmacy had progressed since ancient times and had benefited especially from recent developments in chemistry, it still had need of much improvement, particularly through the removal of some of the useless ingredients of the traditional polypharmacy.[43] Despite Cadet's alarm about the progress of radical simplifying tendencies, there was some hope among the pharmacists that the old polypharmacy could still be saved through reform and rationalization. The Bordeaux pharmacist Chansarel attempted to contribute to such a reform through chemical analysis of the constituents of theriac and a few other composite drugs. The Société's response to his efforts was cautious. While conceding the importance of chemistry in the preparation of medicines, the Sociéte's reporters Athenas and Pierre Alyon pointed out that the physiological effects and therapeutic efficacy of drugs could be assessed only in the course of medical practice, and that the old formula for theriac had stood the test of time for too long to be lightly cast aside.[44]

Whatever reservations some members may have had about reform of the old materia medica, the Société de Pharmacie had by this time become an active center for the chemical analysis of plant substances. The registers of the Société for the period 1805–1815 are filled with reports of such analyses, which were undoubtedly strongly encouraged by the director of the Ecole Vauquelin. The dominant sentiment was summed up by one of Vauquelin's students, Robiquet, who remarked in a report of his own analysis of licorice that "vegetable analysis is perhaps of all branches of chemistry the one that may render the best service to the pharmaceutical art. It alone can enlighten us on the nature of a mass of precious medicines, on the best manner of extracting and conserving them, and often it will even have to guide the physician in the kind of use that he can make of these same medicines."[45] In keeping with the academic model, the Société regularly offered prizes in areas of research it considered important. Among the topics proposed for the year 1809 was the chemical analysis of gums and gum resins, the subject that was soon to be explored in Pelletier's first published research.[46]

Pelletier's early work in the chemical analysis of plant and animal materials also included studies of the compounds of gallic acid with plant substances, an analysis of sarcocolla, the chemical examination of several plant coloring matters such as that derived from olive gum, and a study of the action of nitric acid on the nacreous material of human gallstones (his first collaborative study with Caventou).[47] It was only when he entered into an association with Magendie, however, that Pelletier's interests were focused directly on the chemical analysis of medically or physiologically active plant substances. The first concrete result of this association was a memoir read to the Academy of Science on 24 February 1817 entitled "Recherches chimiques et physiologiques sur l'ipécacuanha."[48] The interest of this memoir was threefold: chemical, physiological, and medical. The primary aim of the chemical section was to isolate the active principle of ipecacuanha bark and to determine its chemical properties. The mean of several analyses of the bark of *psycothria emetica* gave the following proportions of constituents: fatty and oily matter, 2; vomitive matter, 16; wax, 6; gum, 10; starch, 42; ligneous, 20; and loss, 4 (total = 100). Pelletier and Magendie proposed the name *emetine* for the vomitive matter, basing their choice on its physiological effect and on the name of the plant from which it was extracted.

Physiological tests of the products of analysis of ipecacuanha were restricted to the fatty matter and the emetine. Pelletier and Magendie found the former to be the carrier of the strong, disagreeable odor and taste of the crude medicine, but that it was entirely lacking in vomitive action. They found emetine to be odorless and tasteless, but a half grain produced vomiting in a cat followed by several hours of sleep and a return to health. They obtained the same results with dogs and several other cats, and with themselves when they took similar doses. They administered a high dose, twelve grains, by mouth to a dog. Vomiting, followed by sleep, occurred after half an hour, and the dog died fifteen hours after administration. Autopsy revealed a violent inflammation of the lung tissue and of the mucous membrane of the intestinal canal from cardia to anus, a result which Magendie compared to that obtained in his earlier independent study of emetic poisoning. The same results ensued when they gave other animals six grains of emetine, and when they injected emetine dissolved in water into the jugular vein, pleura, anus, or muscular tissue. Pelletier and Magendie pointed out that this localized action of emetine justified its use in some specific illnesses. Clinical trials showed that emetine could replace ipecacuanha as a medicine in all instances without the disagreeable taste or odor of the latter and with greater certainty and constancy in its effects.

Pelletier and Magendie's paper on ipecacuanha already embodied a number of features that were to be characteristic of the Pelletier-Caventou research on the alkaloids. These included choice of a plant substance known to be medically or physiologically active; concentration on isolation of the active principle and intensive study of its physical and chemical properties; use of physiological tests both to aid in the identification of the active principle and to study its physiological mode of action; and resort to clinical trials to compare the effects of the active principle with those of the medicine in crude form. This was not yet a part of a general research program for the isolation of active principles of plant drugs, but merely a statement of one particular result. Although they noted the alkaline properties of emetine and its reactions with acids to form salts, they placed no special emphasis on these results and did not identify emetine as a plant alkali.

The transition from the restricted goals of the joint Pelletier-Magendie paper to the general research program for the discovery of salifiable plant bases subsequently carried out by Pelletier and Caventou occurred in the same year, 1817. This followed publication of a French translation of a paper by the German pharmacist, Friedrich Wilhelm Sertürner.[49] Sertürner reported the successful isolation from opium of the active principle, called morphium after the god of dreams, and of a new acid which he termed meconic acid (*Mekonsäure*). He identified morphium, obtained in crystalline form, as a salifiable base. Though the greatest novelty of his work lay in his discovery of a *plant* base, he did not underline this aspect of his findings and it is likely that he did not appreciate its full importance. Sertürner's editor, Ludwig Wilhelm Gilbert, did recognize the novelty of his claims but thought they were wrong. For Gilbert, plant principles and salifiable bases were mutually exclusive categories, and morphium belonged in the former. Neither Sertürner nor Gilbert recognized the need for a new chemical category.

The first to do so were the French chemists. Commenting on the translation of Sertürner's 1817 paper that appeared in the *Annales de chimie et de physique,* the editor, Gay-Lussac, declared that the discovery of morphine would open a new field, especially for the study of poisons drawn from plants or animals. Already known for their nitrogenous nature and alkaline properties, these substances would "henceforth form a genus of which the species will be found in very different plants."[50] The pharmacist and chemist Robiquet seconded Gay-Lussac's views, praising Surtürner for having "the ingenious idea of considering morphine, contrary to all analogy, as an alkaline substance," and predicting that this idea would "necessarily lead to more exact notions on plant and animal poisons."[51]

Pelletier and Caventou turned Gay-Lussac's prediction into a concrete research program. Sometime in 1817, following publication of the French translation of Sertürner's paper, the two pharmacist-chemists, undoubtedly encouraged by a continuing association with Magendie, turned their attention to the *strychnos* family of plant poisons. The first results of their efforts were read to the Academy of Science in December 1818, and later published in a memoir that well exemplifies their work on the plant bases.[52]

The problem was first posed in a botanical framework. Linnaeus had thought that plants of the same family or genus were most often endowed with analogous medical properties, a view supported by most later botanists. If this was the case, was it not because plants of the same family were composed of the same immediate principles, including one with a special action on the animal economy? Pelletier and Caventou were now prepared to formulate such a proposition in general terms. The medical properties of plants were to be understood as a result of the nature and quantity of certain immediate principles, and these were usually characteristic within the same family of plants. "It is with the goal of establishing these truths in an incontestable manner," they continued, "that we have undertaken chemical research on the most active plants of the materia medica."[53]

Whether by caution or oversight, this initial formulation of their program lacked an explicit statement that the active principles of plant drugs should be all or mainly alkalis, although later in the memoir it became evident that Pelletier and Caventou definitely had this in mind from the beginning of their research on the *strychnos* family. The reference to the materia medica linked the project with the ongoing concern among pharmacists and physicians for the rationalization of drugs and drug therapy. Most important, however, was the general and abstract character of the statement: it was not simply a summing up of previous results, but an anticipation of what was to be expected, and therefore a program for research.

Among groups of plants sharing analogous properties, Pelletier and Caventou continued, there were several species of the genus *strychnos,* notably *nux vomica,* and St. Ignatius bean. In an oblique reference to the work of Magendie and Delille, they noted that these plants had recently drawn the attention of physiologists and had given rise to "learned dissertations read in the heart of the Academy." Chemical studies of *nux vomica* had been undertaken by Henri Desportes and also by Henri Braconnot without isolation of the active principles. St. Ignatius bean had not been studied chemically before Pelletier and Caventou took it up, and it was by turning to this analysis that they had succeeded in isolating the active principle of the poisonous *strychnos* plants as a group. The substance isolated from St. Ignatius bean was crystalline, perfectly white, and "endowed with the distinctive and characteristic properties of the salifiable bases—that is, the faculty of uniting with acids, and saturating them by forming with them true salts that are neutral, soluble, transparent, and crystallizable."[54] The active principle of the *strychnos* plants was, then, a salifiable organic base, like morphine. Pelletier and Caventou proposed to call it strychnine, after its botanical family. Their memoir described its isolation and its chemical and physiological properties.

The directed character of Pelletier and Caventou's research was clearly reflected even in their relatively straightforward account of laboratory procedures. They concentrated on isolation of the active principle, and other aspects of the analysis played only a subordinate role. They presumed from the outset that they would find only a single, chemically uniform active principle, and that this would be the same for all members of the *strychnos* family.

Finally, their haste to examine the white crystalline principle and their application of appropriate tests indicated that they had at least a strong suspicion that it would be alkaline.

Having succeeded in isolating strychnine, Pelletier and Caventou narrowed their investigation to a study of its properties.[55] After a wide-ranging exploration of the chemical properties of strychnine, they resumed the proximate analysis of St. Ignatius bean and reported its constituents to be (1) igasurate of strychnine, (2) a small quantity of wax, (3) a solid (*concrète*) oil, (4) a yellow coloring matter, (5) gum, (6) starch, (7) bassorin, and (8) plant fiber. Exact quantities were not determined, but it was noted that *nux vomica*, analyzed in the same way, furnished the same products, though in different proportions: less salt of strychnine, a greater quantity of solid oil and coloring matter.[56]

In their report on the chemical analysis of St. Ignatius bean, Pelletier and Caventou had already mentioned their use of physiological properties as analytical tools, and the actions of pure strychnine on the animal economy had been joined to its physical and chemical characteristics in the list of its properties. In the final section of the memoir, they reported their physiological findings in more detail.[57] Noting that Magendie and Delille had already determined the mode of action of *nux vomica* on the animal economy, Pelletier and Caventou pointed out that their work had left unresolved the problem of which of its constituents were responsible for its physiological activity. This question, "of equal interest to the physiologist and the chemist," could now be answered on the basis of the preceding chemical analysis. A half grain of pure strychnine taken from St. Ignatius bean was placed in the throat of a rabbit. Convulsions followed after two minutes, and death occurred in five minutes in an attack of tetanus. A half grain of pure strychnine was placed in an incision made in the back of a rabbit. Tetanus occurred after sixty seconds, and death in three and a half minutes. The same experiments, repeated with pure strychnine drawn from *nux vomica*, produced the same results. Strychnine and morphine were administered together, to determine whether the effects of the latter might neutralize those of the former. The mixed results of these trials indicated that morphine or opium extract might ameliorate but not prevent the effects of strychnine. Pelletier and Caventou appended to their results a note from Magendie in which the latter confirmed that the effects of pure strychnine on the animal economy were the same as those earlier observed by himself and Delille for *upas tieuté, nux vomica,* and St. Ignatius bean. Magendie further observed that the action of pure strychnine and of several of its salts appeared to be even stronger than that of the alcoholic extract of the *strychnos* plants, and reported a clinical case in which he had successfully replaced the alcoholic extract of *nux vomica* with pure strychnine.

Surveying the results of these and other experiments, Pelletier and Caventou concluded that in the three plants of the *strychnos* family investigated, the salifiable base strychnine was the sole active principle, and that the salts of strychnine were more active than the base alone. No chemical combination of strychnine would render it nonpoisonous, and therefore the only substances that might act as remedies were those with independent and countervailing actions on the animal economy such as emetics, opium, or morphine. They noted that the spasmodic movements leading to asphyxia in strychnine poisoning might also be prevented by "the surgical operations and the mechanical means indicated by the physiologists whom we have already cited."[58]

These conclusions reflect the several facets—botanical, chemical, physiological, and medical—of the two pharmacists' interest in the *strychnos* poisons. Understandably, the chemical interest was dominant, for this was where Pelletier and Caventou made their own contribution. Though their problem was initially posed in botanical terms, it is likely—given the predominantly chemical and physiological background to their project—that this was something of an afterthought, though an appropriate one.

The physiological and medical dimension of their work was somewhat rough, generally reducing to two questions: Did a given extract or its derivatives have the externally observable physiological effects of the substance in combination? And how did the energy of these gross effects vary with the size of the dose of the pure drug? They were little concerned with assessing the working of a drug within the organism, its precise site or mode of action. In the case of strychnine, of course, Magendie and Delille had already done much of that work. But Pelletier and Caventou also realized that they lacked the techniques and skills of the "practiced physiologist" that would enable them to carry their investigation below the surface of the organism. It was only their continuing collaboration with Magendie that allowed them to confirm the more precisely defined action of strychnine on the animal economy.

Yet their interest in the physiological effects of the substances they isolated and their openness to the results of physiological analysis contrasted sharply with Sertürner's curt dismissal of animal experiment, and reflected the rising visibility and prestige of experimental physiology in France in the second decade of the century. Moreover, their use of physiological criteria, however rough, in chemical analysis, made of physiology a tool science of chemistry, an inversion of the more usual relationship. In further research on the plant poisons, physiology would be more than a merely passive beneficiary of advances in chemical analysis. With their first paper on strychnine, Pelletier and Caventou had given a decisive beginning to that research, which would soon bear out the predictions of Gay-Lussac and Robiquet.

In the years immediately following their strychnine work, Pelletier and Caventou isolated a series of other salifiable plant bases, including brucine, veratrine, cinchonine, and quinine. These investigations shared characteristics that had first been embodied in the research on the *strychnos* plants, including the choice of a highly active plant drug as object of analysis; concentration on the isolation and chemical study of the active principle; reliance on proximate analysis combined with simple animal experiments; a firm expectation that they would find the active principle to be a salifiable organic base; a medical interest expressed in a call for clinical trials or speculation on their probable results; and a continuing association with Magendie.[59] In 1824 they completed their study of the *strychnos* family by carrying out chemical analyses of *upas tieuté* and *upas anthiar*.[60] Obtaining samples from the Muséum (labeled by Leschenault) and from Magendie, they presented their chemical analysis as a needed complement to Leschenault's botanical study and Magendie's physiological study of the *upas* species. Successfully isolating strychnine from *upas tieuté,* they found that the relative proportions of strychnine and brucine—the latter was much less active physiologically—in *nux vomica,* St. Ignatius bean, and *upas* corresponded to the relative strengths of the physiological actions of these substances, and therefore that the chemical and physiological results were in agreement.

Although there was a medical dimension to all of Pelletier and Caventou's work on the plant drugs, it was their analysis of various cinchona barks that resulted in the most tangible benefits for medical practice. These analyses were conceived and carried out as a further extension of the program presented in the strychnine paper. Pelletier and Caventou isolated not one but two organic bases, cinchonine and quinine, in the cinchona barks. Confronted with the objection that the well-known febrifuge qualities of cinchona bark must be due to the special union of all its constituents, the two pharmacists strongly reaffirmed their commitment to the potential practical value of proximate analysis. Without denying that the other substances joined to the active principle in a medicinal plant might usefully modify its action, they insisted on the value of knowing its properties in isolation. To deny this value would be "to banish the chemical sciences from the sanctuary of medicine," and to lessen the possibility of controlling the mode of therapeutic administration of medicines. The point was crucial, for the whole program of chemical reform of the materia medica and of drug therapy was at stake. Only precise knowledge of the active principle "clarifies the pharmaceutical preparations of medicines, makes known rational formulas, and distinguishes them from those that are empirical, absurd, and often dangerous."[61]

Words became actions as Pelletier and Caventou undertook a rational examination of the main pharmaceutical preparations having cinchona as a base.[62] The result was a critique of conventional preparations and a suggestion of new ones based on the isolated active principles. The two pharmacists pointed out that the sulfate and acetate salts of quinine and cinchonine should be especially effective as medicine, since they were soluble without being deliquescent, whereas pure cinchonine and quinine were very insoluble. This prediction was borne out in clinical trials undertaken by the physician François Double (Pelletier's brother-in-law) and others.[63] By the mid-1820s the therapeutic value of sulfate of quinine in the treatment of fevers was firmly established, and its manufacture by Pelletier and others had become the basis of an industry involving substantial commercial interests. In 1827 Pelletier and Caventou, who could by then argue that their isolation of quinine was an important advance for both chemistry and medicine, were awarded the Montyon prize of ten thousand francs by the Academy of Science.[64]

In making this award, the Academy in effect honored not only Pelletier and Caventou but also a whole generation of pharmacist-chemists who had answered Fourcroy's call to make the analysis of plant and animal substances their special preserve, a generation of which the discoverers of quinine were only the leading members. The new experimental physiology, with Magendie as its foremost representative, had played a significant role in the most recent, successful phase of their investigations. Now, through those investigations, physiologists and physicians found themselves in possession of chemical tools of unprecedented precision and power.

Already rather numerous, these substances act in a small dose, and they are not mixed with any principle that masks or hinders their action. Their effects are clear and unmistakable, since they have been carefully studied on animals and on the healthy and sick man. Since their chemical properties are known and the procedures that created them perfectly defined, there is no reason to fear variations in their strength or mode of action. Each of them presents us with a medicine in its greatest simplicity, but also in its greatest energy.

—*François Magendie* (1821)

7 Experimental Pharmacology

In the three decades after 1810 the new physiology allied with pharmacy to produce the first tentative beginnings of experimental pharmacology. The field was directed from its inception to the dual aim of the rationalization of drug therapy and the use of drugs to elucidate human and animal function. Both the use of chemistry in physiology and directions taken by research were strongly conditioned by the problems, opportunities, and support afforded by the Paris institutional environment and the current state of medicine and the sciences. With its encouragement of experimentalism, the First Class of the Institute also urged upon physiologists the example and knowledge of the physical sciences. Skepticism within the medical community about the existing materia medica and drug therapy opened the way to reform. Cultivation of proximate organic analysis by a scientifically ambitious pharmacy profession provided the materials for rationalization in a lengthening list of purified active principles of medicines. The centering of medicine in the hospitals supplied professional positions for physiologists and a field for the testing of new medicines. The increasing prevalence in Paris medicine of the localism, anatomical concreteness, and systematic use of autopsy characteristic of pathological anatomy ensured that the same qualities would mark experimental studies of drug action once they were undertaken. In this context, experimental pharmacology was born. Its advent was signaled by sporadic protests from those for whom the association of chemistry and experimental physiology with medicine was untenable or threatening. Such resistance, which had been endemic to the Paris medical community for decades, could limit but could not prevent a role for the sciences within medicine.

The principal architect of the new experimental pharmacology was François Magendie. Beginning in the 1810s a chemical layer was superimposed on the older, operative and anatomical stratum of his physiology. The physiological and clinical action of drugs came to be one of the two or three major foci of his research. As the operative and anatomical layer of his physiology was indebted to the surgical qualities of his training and experience, so were his drug action studies in part the product of his associations with pharmacists. The isolation of pure active principles of medicines by pharmacists exactly complemented Magendie's project to elucidate the action of those principles in the body. A therapeutic skeptic not satisfied with the passivity usually imposed by that posture, Magendie was driven in part by the desire to reconstruct drug therapy on a sound scientific basis. His entry into private medical practice in 1813, and his increasing association with the Paris hospitals after 1818, gave practical urgency to that desire and opened the way for clinical trials of new drugs.[1] His results were summarized for practicing physicians in the *Formulaire pour la préparation et l'emploi de plusieurs nouveaux médicamens,* which went through nine editions between 1821 and 1836.[2]

The way to Magendie's use of the new drugs had been prepared by his earlier research on the physiological action of the *strychnos* plant poisons and other studies. By the time the first alkaloids were isolated in 1817–1820, Magendie had already gone beyond his initial *strychnos* studies to investigate the physiological action and clinical utility of emetic (tartrate of antimony and tartrate of potassium) and of prussic acid, and had carried out his joint study of emetine with Pelletier. He had followed closely the progress of contemporary analytical chemistry for a number of years and had developed a personal relationship with the chemist Michel Chevreul, who was just then beginning his fundamental research

in organic analysis. Apart from his drug action studies, Magendie
had given substance to his commitment to join chemistry and
physiology through a collaborative study of intestinal gases carried
out with Chevreul, and through his own experimental investigation
of the nutritive value of nonnitrogenous substances.

Magendie's early research on drug action was carried out against
a background of increasing skepticism in the Paris medical community regarding the value and efficacy of the traditional drug
therapy and materia medica. Influential physicians such as Philippe
Pinel counseled a cautious, expectant approach to therapy and a
minimal use of drugs.[3] Partly as a result of Pinel's encouragement,
a number of prominent physicians, including Desbois de Rochefort, Bichat, Schwilgué, Alibert, and Barbier, launched attacks on
the traditional materia medica and undertook projects for its reform.[4]

Despite superficial variations in approach, the positive programs
of these reformers of materia medica shared a number of fundamental characteristics that made them members of the same class.
Most striking was their holistic character: each attempted to embrace all of materia medica or drug therapy in a single, comprehensive system. Related to this aim was a consistent emphasis on
the importance of an adequate classificatory scheme and system
of nomenclature. In this sense, these projects still bore the stamp
of the eighteenth-century penchant for classification, embodied in
France in the analytical program. The hopes placed in taxonomy
were less surprising in view of the active interest in reform of
drug therapy taken by Philippe Pinel, whose *Nosographie philosophique* had appeared in 1797 and served as a standard text in
pathology in the first decade of the century. Also prominent was
the stress on the need for systematic clinical experience and, explicitly or implicitly, on the central role of the hospital.

These emphases reflected the circumstances of the birth of programs for reform in the context of the new, hospital-based medicine. From the perspective of this medical milieu, there was little
reason to expect the appearance of a laboratory or experimental
approach to the study of drug action, and it is hardly surprising
that "physiology," as used in the context of these reform programs,
had primarily the sense of Bichat's vital-properties doctrine rather
than of the experimental-functional approach represented by Bichat's studies of violent death or by Magendie's early work. In

spite of their occasionally contemptuous remarks about the traditional drug therapy, none of these reformers was prepared for a truly radical break with the past, and all were ready to retain a large part of the traditional materia medica, at least pending further investigation.

An entirely different approach to the problem of drug action emerged from the early physiological work of François Magendie. Magendie had not escaped the prevailing current of therapeutic skepticism. On the contrary, he was probably the most thoroughgoing of the skeptics among Paris physicians. By nature scornful of systems whether in physiology or therapeutics, he bluntly dismissed all deductive approaches to therapy. When cholera struck Paris in 1832, he noted without surprise the impotence of treatments derived from the pet theories of physicians.[5] Traditional polypharmacy—"these learned prescriptions to which all the kingdoms of nature contribute, and which, dictated with a mysterious dignity, seem rather made to raise the merit of the physician in the eyes of the common man than really formulated in the interests of a cure"—fared no better in his judgment. In such a mood he was liable to counsel a passive or expectant approach, preferring to prescribe nothing rather than "to throw out at random a prescription the effects of which will be what they may."[6] On the whole, however, he saw the value of doing something harmless, if physiologically ineffective, to put the patient's mind at ease and satisfy the conscience of the physician.[7] Even after he was well advanced in the development of the new chemical therapy, he remained cautious, noting that physicians were too prone to attribute cures to the use of specific medicines, whereas in many cases equally good results could be had from a number of treatments, including simple bed rest and mild drinks.[8]

By its nature, therapeutic skepticism could provide only the negative conditions for a reform of drug therapy. Positive direction for reform had to come from elsewhere. In Magendie's case, it came initially not from the clinic or from the classificatory ambitions of the analytic program but from problems that emerged naturally in the course of his early research in experimental physiology. In his early studies of the physiological mode of action of the *strychnos* poisons, Magendie had arrived at the important result that these poisons produced their observed effects through direct action on the spinal cord following absorption into the circulatory system. His next investigation of drug action came about as an unanticipated development of experimental studies on the mechanism of vomiting.

On 1 March 1813 Magendie read to the First Class a memoir, "Sur le vomissement," in which he attempted to demonstrate, against generally received opinion, that the stomach played a passive role in vomiting and that the principal action was performed by the diaphragm with some auxiliary aid from the abdominal muscles.[9] In the opinion of the distinguished panel of commissioners named to judge this memoir, he had succeeded so well in his aim as to have placed the question beyond further doubt. The commissioners largely credited this outcome to the fact that although Magendie had made use of an old and well-known medicine in his investigation, he had studied it not with reference to medical practice but rather "as a skillful physiologist and judicious experimenter."[10]

While fully confirming the particular result, the commissioners attempted to place it in a more generalized setting and so evoke questions for further research. Magendie's experiments underlined, first of all, the dependence of the act of vomiting on the nervous system, and therefore made this act a special instance of the more general class of phenomena characterized by Legallois.[11] The precise means by which an emetic introduced into the stomach affected the nervous system remained to be determined. There were two alternatives: either a direct irritation of the nerves of the stomach or absorption into the circulation followed by transport to the brain or spinal cord. The first alternative appeared to be favored by experiments in which sectioning the eighth pair of nerves was followed by vomiting, the second by Magendie's experiments in which vomiting was provoked when an emetic was injected into the bloodstream of animals whose stomachs had been replaced by passive bladders. In support of the latter alternative, the commissioners referred to Magendie and Delille's study of the action of *upas,* which had shown that that poison acted on the spinal cord after absorption into the bloodstream. They went on to remark that very probably "almost all substances that have some effect on the animal economy act in this manner" and that this opinion "leads us to views entirely new respecting the mode of action of most medicines and poisons."[12] Greatly impressed with Legallois's and Magendie's research, the commissioners of the First

Class were now prepared to venture a generalization that Magendie himself would probably have regarded as premature.

One further question remained. Legallois had shown that the muscular movements of respiration were dependent on the portion of the medulla oblongata giving rise to the eighth pair of nerves. A principal task of physiology would now be to localize the seats of other functions in the brain and spinal cord. Where, in particular, was the seat of vomiting? The commissioners pointed out that certain experiments of Legallois suggested that it might be very near to the one controlling the muscles of respiration.[13]

Magendie read his response to the commissioners' report in August.[14] He pointed out that although emetic (tartrate of antimony and tartrate of potassium) had been well studied in its chemical composition and clinical uses, its physiological mode of action remained unknown. Clinical experience and experiments on over fifty dogs and cats showed that a large dose of emetic could be taken by mouth without fatal consequences if most of it was vomited up. To get at the main question—the mode of action of emetic—Magendie followed the procedure of his *strychnos* poison experiments and placed a known quantity of emetic in contact with a variety of absorbant surfaces, including the mucous membranes of the large and small intestines, the various serous membranes, the peritonea, the pleura, cellular (connective) tissue, and the tissues of some organs. In all but one case vomiting ensued, followed or preceded by excretion of fecal matter. The single exception was the pleura, for which application of emetic in more than twenty separate experiments never produced vomiting, and only rarely evacuations. Venous injection of emetic produced the same symptoms more quickly and with greater intensity. With moderately large doses (for example, four to eight grains) other symptoms, including disturbed respiration, appeared and death occurred after several hours. Autopsy revealed extensive alterations of the lungs and mucous membrane of the intestinal canal. When emetic was administered by mouth and the esophagus tied to prevent vomiting, the same kinds of effects were produced, but much more slowly. Magendie concluded that emetic probably acted after absorption rather than by some direct effect on the stomach. He underlined the implications for medical practice: the utmost caution should be exercised in applying emetic as a friction or lotion, and venous injection must be considered dangerous. Finally, he reported experiments showing that section of one or both pneumogastric nerves dramatically slowed the action of the drug.

While there is no doubt that the paper on emetic was a direct response to the recommendations expressed by the commissioners of the First Class, there was a disproportion between the expectations of their report and Magendie's conclusions. Citing the "general principle" of Legallois, the commissioners had placed great emphasis on the probable role of the nervous system in controlling the body's response to emetic, and had even speculated on the precise location of the seat of vomiting in the spinal cord. Magendie, in contrast, nowhere referred to the ultimate effect of emetic on the brain or spinal cord, and he did not speak of emetic acting on the nerves of the stomach. The emphasis of the commissioners' report on the controlling role of the nervous system remained, at best, implicit in his article. That Magendie had it in mind is shown by his experiments on the effects of sectioning the pneumogastric nerves, but even here his conclusions scarcely went beyond a bare restatement of the facts. Here one sees Magendie's characteristic caution in drawing abstract or general conclusions, even where his experimental work was guided by a commitment to their probable validity. Despite Magendie's failure to address directly the questions on the role of the nervous system posed by the commissioners, the report of the First Class was entirely favorable. All of Magendie's experiments were repeated before one of the commissioners, and their results confirmed; and the four commissioners of the memoir on vomiting gave their unqualified approval to its sequel.[15]

The origin of Magendie's study of emetic in his prior experimental work on the mechanism of vomiting was partially disguised by introductory remarks that emphasized the medical history and utility of the salt, by citations of clinical cases that illustrated its gross effects, and by conclusions drawn regarding its mode of administration to patients. The medical emphasis is not surprising in view of Magendie's decision to begin a private practice in the same year (1813), and no doubt that decision lent a new urgency to his drug action studies. Nevertheless, as with the investigation of the *strychnos* poisons, the physiological interest predominated. Magendie did go beyond the approach of the 1809 memoirs in his systematic use of autopsy to determine with precision the lesions induced by the poison. In so doing, he gave clearer expression to the deep parallel between the localism and anatomical concreteness of his own experimental approach to the study of drug action and the same qualities in the pathological anatomy then gaining influence in Paris medical circles. Even here, however, Magendie adapted pathological anatomy to the needs of an active, interventionist experimental physiology, making it more than an adjunct of clinical medicine. Magendie had serious reservations about the value of a pathological anatomy unaided by physiology, because the lesions it studied sometimes appeared only after death. In his view, these lesions were merely the effects of disease— effects that could be controlled only by improved knowledge of causes.[16]

Magendie's study of the mode of action of emetic emerged naturally from his early work in experimental physiology, with significant encouragement from the commissioners of the First Class. Though it fully embodied the anatomical, surgical, and operative qualities of that physiology, it did not yet reflect Magendie's interest in chemistry or his concern to turn recent advances in chemical knowledge and technique to the advantage of physiology. That interest and concern developed in the two or three years between the appearance of the paper on emetic and the publication of the first edition of the *Précis élémentaire de physiologie* in 1816–1817.[17]

Digestion and nutrition provided Magendie with his first important investigations on the common ground between physiology and chemistry. The first of these, done in collaboration with the young analytical chemist Chevreul, was a study of the intestinal gases of the healthy man.[18] A paper on the nutritive properties of nitrogen-free aliments, which was read to the Academy of Science on 19 August 1816, joined chemistry and physiology by focusing attention on the chemical composition of ingested and excreted matters and secretions, and by insistence on exact qualitative analysis of these substances.[19] In particular, by carefully excluding nitrogen-containing substances from the ingested material of experimental animals, Magendie was able to give a highly creditable answer to the important question of the origin of the nitrogen of animal tissue. Magendie concluded that the nitrogen of animal tissue probably came from ingested aliments.

A distinct but closely related emphasis on the chemical purity of substances introduced into the organism in experimental investigations was to be a characteristic feature of Magendie's studies of drug action. On 24 February 1817, six months after the appearance of the memoir on nitrogen-free diets, Magendie and Pelletier presented to the Academy of Science the results of their joint research on ipecacuanha. In this investigation, chemical isolation of the active principle in pure form was both a major goal and the prerequisite for study of the physiological effects and rational medical use of the medicine, hitherto available only in combination. The same emphasis on the chemical purity of an isolated active principle and its use in physiological and clinical studies was present in Magendie's memoir on the role of prussic or hydrocyanic acid in the treatment of diseases of the chest, published later the same year.[20]

The memoir on prussic acid opened with a general statement underlining the "utmost importance" of physiological experiments for the practice of medicine. It was by means of such experiments, Magendie maintained, that the really active remedies were separated from those used as medicines "on mere hypothetical principles," and that new remedies were discovered.[21] With these ends in view he had undertaken to investigate the poisonous and medicinal qualities of *upas tieuté, nux vomica,* emetic, and ipecacuanha. He would now present the results of a similar study of prussic or hydrocyanic acid.

The external symptoms and physiological effects produced by prussic acid were, in fact, fairly well determined before Magendie turned his attention to the subject.[22] From the work of earlier authors and some experiments of his own he drew several general conclusions—that prussic acid, whether in liquid or vapor form, was injurious to the life of all animals and many plants; that death was produced more rapidly in proportion to the rate of circulation and the capacity of the respiratory organs; that prussic acid acted on warm-blooded animals by destroying their sensibility and the contractility of the voluntary muscles; and that the same happened to man when the dosage was sufficiently high.

Recently, Gay-Lussac had made known a new process for the preparation of prussic acid which yielded that substance in pure form, undiluted by water.[23] If the dilute form had been a strong poison, Magendie noted, the effects of the acid in pure form were truly frightful. This was illustrated by three simple experiments.

In the first, the extremity of a glass tube previously dipped in pure prussic acid was inserted into the throat of a strong dog. Scarcely had the tube come in contact with the tongue when the dog made two or three long and rapid inspirations and fell dead in a state of rigidity. A subsequent examination of the locomotor muscles— of which Magendie did not give any details—revealed no trace of irritability. In a second experiment, in which "a few atoms" of the pure acid were placed on the eye of a strong dog, the effects were almost as rapid as in the first case, and otherwise the same. In a third experiment, one drop of the pure acid diluted with four drops of alcohol was injected into the jugular vein of a dog, and the animal fell dead at that instant, "as if struck by a cannon shot or by lightning."

Magendie concluded that the pure prussic acid, prepared according to the method of Gay-Lussac, was the most active and rapidly fatal of the known poisons. Nevertheless, in Magendie's view there was no doubt that when properly diluted in water it could safely be used as a medicine.

In studying prussic acid poisoning, Magendie had often observed that animals which had lost all trace of sensibility and of locomotor muscle contractility could continue to breathe for several hours with intact though accelerated circulation. They could be said to be dead with respect to their external functions, while still alive in their nutritive faculties. These experimental results suggested a possible direction for clinical trials: prussic acid might be used with advantage in treatment of diseases due in part to an excess of sensibility.[24] Having proceeded on this view, Magendie was able to report success in the use of dilute prussic acid in the treatment of cases of nervous and chronic cough that had come to his attention in the course of his private practice.

With these encouraging results in hand, Magendie was led to ask whether prussic acid might alleviate the cough and other symptoms of pulmonary consumption, or even retard or arrest the disease. In treating over a three-year period fifteen patients affected with phthisis, he had succeeded in relieving the symptoms of the disease. When prussic acid was given in small but repeated doses, coughing was diminished, expectoration was moderated and facilitated, and sleep was made possible without the sweats associated with the use of opiates. The same results had been obtained in clinical trials undertaken since the previous August (1817) by

Théodoric Lerminier at the Charité hospital. At Magendie's instigation, Lerminier had administered prussic acid in a dosage of four drops diluted with water to twenty patients suffering from various stages of phthisis. He had achieved alleviation of symptoms in most cases, and this relief was more marked when the disease was in an incipient phase.

The larger question of whether prussic acid might retard or arrest the progress of pulmonary consumption remained to be answered. Magendie announced that he was continuing his collaborative work to that end with Lerminier at the Charité, where twenty or thirty consumptive patients were ordinarily received for treatment. Such a bold attack on one of the most dreaded diseases of the day was justified, Magendie asserted, by the recent appearance of new substances notable for their powerful physiological action.[25]

When in 1820 the original paper on prussic acid was reprinted in English translation, Magendie was able to add fifty-three pages reporting further clinical experience with the drug. For the most part these pages described the experience of other physicians in France, Britain, Italy, and the United States, reflecting the wide interest aroused by the original paper. Prussic acid had been employed in the treatment of a variety of conditions, with some degree of success. The Paris Faculty of Medicine had placed it on the list of recommended medicines in the new codex. Magendie had continued to concentrate on treatment of phthisis pulmonalis, with generally positive results.[26] He gave a description of the formulas including prussic acid that were most frequently employed in his own practice.

At the close of the 1820 portion of his paper, Magendie remarked that the case of prussic acid showed that physiology, chemistry, and medicine could be effectively allied in the cause of humanity.[27] Although scientific and medical interests had been linked from the beginning in Magendie's work, the medical side was gaining increasing prominence, no doubt in part because of his appointment by the central board of the Paris hospitals in 1818. Whereas in the earliest papers on the action of the *strychnos* poisons and on absorption the primary end had been to determine the nature and sequence of physiological events, the main concern of the prussic acid paper was to identify the possible uses of prussic acid in medical practice. Of the four general conclusions of the original paper, only one referred to the poisonous action of the pure prussic acid, while the other three all referred to its medical uses. More important, the introductory remarks underlined the role of physiological experiment in informing medical practice by defining and improving the efficacy of old and new medicines, reflecting a subtle shift of emphasis from pursuit of physiological knowledge for its scientific content to pursuit of that knowledge as the basis of therapeutics. The same shift of emphasis appears between the nitrogen-free diet experiments and the attempted clinical exploitation of their results in a paper on gall bladder and urinary bladder stones, which occurred in the same period. The original paper on prussic acid appears to have provoked interest overwhelmingly as it related to therapeutics. Some of this response, including the greater part of the additional material presented in the 1820 edition, was favorable. Some, evidently, was not, for Magendie complained in his preface of the prejudice of some physicians against the sciences, "which at this moment declaims against the utility of physiological experiments and the rational and restricted application of chemistry to medicine."

Just such an application of physiology and chemistry to medicine was now one of Magendie's major goals. By 1817 Magendie had had some success in assessing the physiological effects of the *strychnos* poisons, emetic, ipecacuanha, and prussic acid, and had clarified the use of these substances in therapeutics. In the cases of emetine and prussic acid, he either had been directly involved in chemical isolation of the active principle or had drawn on the most recent results of chemical analysis. He was then in a position to formulate a general program, and it was this program, as well as its embodiment in one particular case, that was given expression in the paper on prussic acid. It is probably no coincidence that it was in the same period, in the wake of Magendie and Pelletier's collaborative study on ipecacuanha, that Pelletier and Caventou were beginning their own investigation of the *strychnos* poisons and formulating plans for the systematic chemical analysis of physiologically active plant substances. The two programs were complementary. One aimed at the chemical isolation of pure active principles; the other promised to submit those principles to tests that would determine their physiological mode of action and therapeutic usefulness. Both combined distrust of the old materia medica and its load of "empirical, absurd, and often dangerous" concoctions with confidence that a rational and effective pharmacopoeia could be created through chemical analysis, physiological experiment, and clinical trials.

Such complementarity of goals and methods is understandable in view of the shared background of attitudes toward drug therapy common in contemporary Paris medical circles. More important was the role played by Magendie in turning Pelletier's attention to physiologically active organic substances. Therapeutic skepticism alone could sweep away the debris of the past, but could not create the foundation for a new therapy.[28] What was required was a new basis for optimism and for fresh efforts at positive construction. Materials were at hand in the increasingly powerful techniques of analytical chemistry, the chemical skills and interest in organic analysis of French pharmacists, the development of experimental physiology by Magendie and others, and the clinics of the Paris hospitals. It was Magendie who saw the possibilities and boldly acted on them, bringing the sciences to bear in a program to create a rational and effective drug therapy. The most skeptical of therapeutic skeptics would now be the most active participant in the drive to implement that program.

A first culmination came in 1821, with the publication of Magendie's *Formulaire pour la préparation et l'emploi de plusieurs nouveaux médicamens*. In spite of the opposition of seventeenth-century physicians to the use of emetic in medicine, Magendie noted in the preface, the utility of antimonial preparations had long been recognized. He hoped the same recognition would be accorded to "the new substances that chemistry and physiology have joined in indicating as precious medicines."[29] Progress in materia medica had long been retarded by the absence of techniques for the isolation of the constituents of medicines by chemical analysis, and by the old but still common view that the effects of medicines on animals were entirely different from their effects on man. Now both of these obstacles were removed, one through the recent progress of analytical chemistry, the other through Magendie's own experience of the past ten years in the laboratory and at the patient's bedside. Firmly convinced that the mode of action of medicines was the same for man and animals, Magendie had determined the physiological properties and medical virtues of most of the substances included in the *Formulaire*. Available in chemically pure form and exactly measured doses, these substances were clear and unmistakable in their effects. "Each of them," he noted, "presents us with a medicine in its greatest simplicity, but also in its greatest energy."[30] The qualities of clarity, simplicity, and certainty, characteristic of Magendie's physiological investigations, would now be the hallmark of his pharmacology as well.

Of the twelve substances discussed in the first edition of the *Formulaire,* seven—*nux vomica*/strychnine, morphine, emetine, cinchonine, quinine, veratrine, and prussic acid—had already been the subject of separate publications. Ten were alkaloids, isolated within the past four years. The two nonalkaloids, prussic acid and iodine, also owed their discovery or isolation in pure form to the recent efforts of analytical chemists. Each substance was allotted a separate section, and Magendie made no attempt to impose a general classification. Each section followed a standard order of presentation: a brief note on the discovery of the substance; information on its preparation in a pure or nearly pure state; its action on animals and on the healthy man in different doses; its uses in the treatment of disease; its dosage and mode of administration. The sections on *nux vomica* and strychnine were typical.[31]

Despite Magendie's substantial research on the physiological mode of action of the *strychnos* poisons, only the briefest summary of its results was presented in the *Formulaire.* The main interest of the book lay elsewhere and was clear from the more extended discussions of the action of the drug on the sick person and of its mode of preparation and administration. The same relative emphasis was maintained in the sections on other substances. It could hardly be otherwise in a work designed as a practical handbook for physicians and pharmacists. Admittedly, new physiological information was included. Narcotine was recognized as a separate constituent of opium, and its physiological effects were distinguished from those of morphine.[32] Magendie was prepared to be more specific about the action of prussic acid, maintaining that it calmed the "too lively irritability that has developed in certain organs" and that it could be used to advantage in "all cases in which the irritability of the lungs is harmfully augmented."[33] He was also able to say something about the physiological effects of three of the substances that had not been the subject of separate memoirs: solanine, gentianin, and iodine.[34] In general, however, the *Formulaire* remained a practical work in which physiological knowledge occupied a significant but clearly subordinate place.

Magendie's project to introduce the new chemically isolated substances into therapy and to use chemistry and physiology to inform medicine met with resistance from the medical community. The precise nature, extent, and sources of this resistance are yet to be fully studied, but indications of it may be found in Magendie's writings and elsewhere. In the preface to the first edition of the *Formulaire* he spoke of "the repugnance which many enlightened practitioners experience" ("still experience" in the preface to the seventh edition) in using the new medicines, and hoped that it would soon be dissipated by continued practice.[35] In 1816 his paper on nitrogen-free diets had been harshly attacked by the editor of the *Journal* of the Royal Institution, who had complained among other things of Magendie's "mania for experiments" and had asked that Magendie be censored by "professional men." Magendie had replied that he would gladly submit his work for judgment to those of the same profession, and had pointedly identified these as the English *physiologists,* such as Everard Home, Benjamin Brodie, James Macartney, John Cross, and Wilson Philip.[36] Magendie also had supporters in the English medical community. In 1818 his original paper on prussic acid was published in the Royal Institution's *Journal of Science and the Arts* with a prefatory letter by the physician Augustus Granville, who had witnessed some of his experiments on prussic acid during a visit to Paris, and was glad to attest to his "sagacity and talents."[37]

Another indication of criticism of Magendie's work appeared in an extract of his paper on urinary and gall bladder stones printed in 1818.[38] Magendie's approach, the author observed, "must be singularly displeasing to those who appear to dread, above all, to see the sure path of experiment introduced into medicine." In defending Magendie, the author noted that for a while it was to be feared that "these partisans of a blind empiricism and of vaguer hypotheses were harmful to the progress of science; but in taking their attacks to extremes they have lost all credit." In attacking men of the stature of William Wollaston or Jöns Jacob Berzelius, critics had only undermined themselves.[39] As the mention of the last two names indicates, it was not Magendie's work alone that evoked hostility, but the whole contemporary trend to bring the sciences to bear on medicine.

Others participating in one or another aspect of Magendie's drug action program (such as the medical doctor Bernard Gaspard;

Etienne Serres, physician to the Pitié hospital; and Magendie's collaborator in his early *strychnos* studies Delille) also found it necessary to defend the relevance of experimental physiology to medicine.[40] Magendie's contemporary Orfila, a physician and founder of toxicology, had to work against the prejudice that physicists and chemists were incompetent when it came to studying living things. He argued that chemistry and physics had much to contribute to medicine, and his *Traité de toxicologie générale* of 1815 exploited the discovery that when poisons are associated with organic matters, they cannot be detected by the same chemical tests as when they are in a state of purity.[41]

In 1820 Magendie took the offensive and published one of his strongest and most explicit statements on the necessary role of the sciences in medicine. In the preface to the expanded version of his prussic acid paper, he attacked the singular and dangerous "mania" according to which a physician should ignore anatomy, physiology, physics, chemistry, and the other natural sciences in favor of an exclusive devotion to practice, or "a stupid routine, masked under the name of experience."[42] It may be that Magendie had caricatured the antiscientific currents in contemporary medicine, especially in view of the emphasis placed on normal and pathological anatomy by the Paris clinical school. There is no doubt that those currents existed, however, and that Magendie felt their pressures.[43] In later writings he attacked the insufficient attention given the basic sciences in French medical education, pointed out that pathological anatomy alone could not reveal the causes of disease, and suggested that experimental physiology could play a crucial role in the training of surgeons.[44]

Magendie's efforts to persuade physicians of the value of the sciences were complicated by the vogue for François Broussais's brand of "physiological medicine," which peaked in the 1820s.[45] An aggressive, dogmatic, and colorful personality, the former physician of Napoleon's armies advocated a system in which many illnesses were perceived as the result of inflammations of the alimentary tract. From these "physiological" principles he derived— with a logic not entirely transparent—a heavy therapeutic emphasis on bleeding, especially by leeches. In Magendie's view, the appeal of Broussais's theory to physicians was that of any systematic physiology offering easy and comprehensive answers by attaching itself "to a single idea, from which it deduces to an absurd point a number of forced consequences." He found repeated occasions to criticize this rival claimant to the name of his science.[46]

The prejudices of medical men about the sciences, Magendie felt, would never be destroyed by reasoning alone, but only through the gradual accumulation of "facts and material proofs" over time. It was to that task rather than to polemics that his energies would be directed.

Polemics or not, Magendie gradually acquired a reputation as the leading exponent of a role for the sciences in medicine and of the alignment of physiology with the exact sciences. That reputation is clear from the cautious and qualified but largely favorable account of Magendie's work published by the *Gazette médicale de Paris* in 1830.[47] The author noted that Magendie had led the way to a renewed interest in the application of chemistry and physics to physiology, out of favor in the Paris school since Barthez, Bichat, and Chaussier. The writer noted that Magendie's predilection for chemistry (*chymisme*), much less exclusive and dogmatic than older views, was only perceptible by close attention to the details of his research. Even for those ignorant of his positive research, Magendie's name was now associated with "a certain chemical doctrine" in medicine. Magendie had also become the leading proponent of animal experiment in physiological investigation. In fact he seemed to make physiology entirely dependent on that form of experience. In the author's view, both aspects of Magendie's approach were open to some criticism. In claiming that physiology had to be based on experience, and also that physiology could not be separated from the exact sciences, Magendie was assuming what needed to be demonstrated. But the idea that the laws of physiology were opposed to or at least different from those of the exact sciences was itself based on experience. Moreover, despite his skill in carrying out animal experiments and the undoubted discoveries he had made by that means, the relative value of the approach was still in doubt. Different experimenters often obtained conflicting results, and controversy was not eliminated. More important, although animal experiment aimed to view a natural process, the experimenter had to disturb nature, often to a considerable extent. How far, then, could Magendie's conclusions be trusted? Neither criticism was fatal, however, and the author pointed out that he merely wished to moderate the exaggerated hopes that Magendie seemed to be creating.

Whatever doubt, hesitation, or outright opposition Magendie encountered from elements of the medical community was more than offset by the slow but undeniable successes of his own research and by the participation of a widening circle of physicians and chemists in similar investigations. In the summer of 1820 Magendie gave his first course on the experimental physiology of drug action, emphasizing his conviction that the action of drugs was the same in dogs and humans.[48] He offered a similar course in the winter of 1822–1823. By that time his attention was shifting to the nervous system, and he concentrated on direct application of medicines to various parts of the brain and nerves.[49] Meanwhile he continued clinical trials—for example, using quinine sulfate in treatment of involuntary movements and neuralgia.[50]

Gradually, he was joined by other physicians and pharmacists. Gabriel Andral studied the action of veratrine on animals and the therapeutic properties of strychnine and brucine. Pierre Segalas d'Etchepare examined the effects of belladonna on the eye and the mode of action of *nux vomica,* in the latter case asserting against Magendie that the poison killed not by causing asphyxiation but by a direct effect on the nervous system. Robiquet and Louis Villermé explored the therapeutic use of solution of pure potassium cyanide as a substitute for prussic acid. Gaspard, with many references to earlier work by Bichat, Nysten, and especially Magendie, investigated the effects of introducing various substances into the arteries of animals. [51]

In 1827 the Academy of Science, recognizing the increasing role played by the physical sciences in physiological investigation, voted to name two new commissioners from the sections of physics and chemistry to the panel of judges for the Montyon prize in experimental physiology. The first to be named to these positions were Gay-Lussac and Thenard.[52]

The *Formulaire* itself was well received, passing through eight editions between 1821 and 1834. In the preface to the last edition, Magendie wrote that practice had fully confirmed the value of the substances described in the early editions. What physician, he asked, would now give extract or powder of cinchona bark instead of quinine sulfate, or the old opium preparations in place of morphine? In the difficult study of the action of medicines there was always room for improvement, however, and Magendie reported that he had made full use of his position at the Hôtel-Dieu to reinforce or qualify his earlier conclusions. The results of these further studies are evident in the more extensive discussions of clinical experience.[53] Chemistry had continued to provide a steady flow of new substances to be examined for possible use as therapeutic agents, and many of these had won admission to the *Formulaire*. Among the latter were new alkaloids such as codeine, and halogens, like bromine. Magendie had broadened the scope of his discussion to include the elementary chemical composition of the new medicines. The time had come, he said, "to seek the relationship between the number, proportions, and nature of their elements and their mode of action on the healthy or sick man."[54]

In passing through eight editions the *Formulaire* had grown from 84 to 438 pages, making available to physicians more and better information on a wider variety of new medicines. In the preface to the last edition, Magendie noted that it was his "scientific and friendly relations" that had enabled him to test the latest chemicals for their effects on animals and, ultimately, in the clinic. Examples included codeine, obtained from Robiquet, prussic ether from Théophile Pelouze, narceine from Pelletier, and meconine from J. P. Couerbe. The text reveals the full extent of Magendie's reliance on other investigators and the breadth of the interest that had been generated in the chemical, physiological, and clinical study of drug action by the mid-1830s. More than 75 chemists or pharmacists and over 120 physicians are cited by name, some more than once. A wide range of nationalities, including French, British, German, Swiss, Irish, Italian, American, Portuguese, and Swedish, are represented.[55]

Magendie's own drug action studies of the 1830s continued along the double track he had set for them two decades earlier. One set of interests was physiological, probing the mode of action of drugs on the body and making of them tools for the elucidation of function. The other was therapeutic, seeking safe and effective remedies against specific pathological conditions. In both areas, advance was slower and more limited than the bulky size of the last editions of the *Formulaire* might suggest. The properly physiological studies were restricted not only by the limitations of contemporary organic and physiological chemistry, but also by Magendie's concentration on other topics, especially the nervous system and physical phenomena such as the mechanics of circulation.[56] Advances in therapeutics were constrained by the lack of physiological understanding of disease, and by the need for caution in clinical trials. The appearance of a new disease—cholera—in 1832 presented physicians with challenges for which even traditional therapies could not offer an immediate response.

In his lectures of the mid-1830s on the physical phenomena of life, Magendie occasionally made use of drugs and other chemical agents. For example, he employed alcoholic solution of *nux vomica* to show the physical character of absorption, that absorption takes place in the lungs, that the intestinal mucous membrane absorbs more readily than the epidermis, and that the veins absorb. He made similar use of other substances, including prussic acid, ether, phosphorous, emetic, and sulfuric acid.[57] He was alert to the results of others—noting, for example, Barbier's claim that codeine acts specifically on the ganglionic nerves, especially in the epigastric region. He pointed out that if it could be verified, knowledge of such a specific action would be a great step toward understanding the physiology of the ganglionic system.[58] As part of his research on the blood, he examined how it was affected by a range of substances, including iodized hydriodate of potassium, emetic, cinchonine, quinine sulfate, and decoction of dry digitalis.[59] Old results were confirmed and new lines of investigation opened up, but no new physiological findings of major importance emerged.

Meanwhile disease pursued its course, unmindful of the pace of scientific investigation. In 1832 the great cholera pandemic, making its way westward, struck Paris with force. Within six months more than 18,000 people would be dead.[60] Magendie, by then a staff physician at the Hôtel-Dieu, felt the brunt of the onslaught. Suddenly his ward was flooded with cholera victims, and he reported to his students at the Collège de France that he was engaged in "hand-to-hand combat" with the disease. Between March and August 1832 he treated 594 cholera patients.[61]

Asiatic cholera was unprecedented in Europe. The etiology was unknown, and there was no established therapy. Those who, like Broussais, hastened to apply favorite theories were soon vanquished, unable to supply either cause or cure.[62] Magendie reported that in eight days at the Hôtel-Dieu "all the active means of medicine" were tried and found wanting.[63]

The experience of the cholera epidemic confirmed Magendie in his therapeutic skepticism without causing him to relinquish the aim of establishing rational chemical therapeutics. He arrived at his system of treatment for cholera by instinct rather than by prior knowledge or advice. On his own admission, he carried out few clinical trials. Those he did perform merely revealed the marginal utility of known drugs in the treatment of cholera. He noted that both patients and physicians, including sometimes himself, were too prone to attribute to medicines the effects that followed them in time, whereas those effects might have other causes. Although not simply passive or expectant, his own therapies were simple and cautious.[64]

The obverse of therapeutic caution was a clear-sighted recognition of the limits of existing physiology. Throughout his day-to-day struggle with cholera, Magendie continued to emphasize contemporary ignorance and the need for better knowledge of the causes and effects of the disease. For that, neither deductive systems à la Broussais nor pathological anatomy as conventionally practiced would suffice. Only thorough study of the disease in the smallest details of its course and its effect on functions could lead to understanding. His findings were regularly presented to classes at the Collège de France; though with bacteriology decades in the future, the cause of the disease eluded him, and the net effect of his physiological results on therapy was small.[65]

Physiology might inform knowledge of disease and its treatment; study of disease might also inform physiology. In 1832 Magendie noted that the facts of pathology observed in cholera were, "so to speak, experiments made deliberately to show the mechanism of different functions."[66] His recognition of the opportunities presented to science by disease was in part due to the pervasiveness of pathological anatomy in Paris medicine by the 1820s, in part a consequence of his own deepening involvement in hospital medicine and its attendant clinical and pathological experience. By the time of the cholera epidemic, Magendie had already been applying his insight to study of the nervous system for over a decade.

What we do not dare to do on man, nature, a less scrupulous experimenter, takes it upon itself to do. How many diseases of the brain and spinal cord are only the faithful reproduction of our experiments! The pathology of the nervous system is thus nothing other than experimental physiology applied to men.

—*François Magendie* (1839)

8 Pathological Physiology

In the 1820s Magendie and other Paris physicians conceived and put into practice an approach to the study of human function that aimed to exploit fully the physiological insights offered by disease. Pathological physiology, as the method was called in the title of Magendie's *Journal* and elsewhere, considered the lesions and functional disturbances observed in clinic and autopsy room as so many experiments on the human organism, experiments that could be profoundly revealing of the nature of the normal processes they disrupted. In essence the idea was not new. The Montpellier school, and Barthez most emphatically, had often urged the observation of the sick person as the key to knowledge of human function. Surgical experience had long been a source of physiological insight, and this was notably the case for Bichat and others of his generation. Magendie, probably echoing Barthez, had underlined the interdependence of pathology and physiology as an ideal in his M.D. thesis of 1808.[1] It was only after 1820, however, that the idea was fully articulated and made the basis of a positive research program embraced by a part of the Paris medical community.

166

Magendie played a leading but by no means exclusive role in the execution of this program, as editor of what was for ten years its primary vehicle of publication, and through his own research on the nervous system. The latter claimed a major share of his attention in the 1820s and again in the period 1838–1842. It had a number of sources, including the comparative anatomy of Cuvier and the Muséum, widespread scientific and popular interest in the contemporary research of Franz Joseph Gall and Johann Spurzheim, and the stimulus of foreign work such as that of Charles Bell, as well as Magendie's own experimental investigations and clinical experience. Pathological physiology joined to the operative or surgical aspect of Magendie's experimentalism dominated the research, while comparative studies played a subordinate and ancillary role.

Close study of Magendie's work on the nervous system reveals that he was not the simple-minded empiricist—the "rag-picker of science"—that he sometimes claimed to be and that later generations, beginning with Claude Bernard, confirmed as his image. Not only did theoretical expectations often play a role in his investigations, as we would expect, but Magendie himself was increasingly, if episodically, aware of this fact in the later decades of his career.

The program of pathological physiology, and the way in which it was developed over more than two decades in Magendie's research on the nervous system, gives evidence of a more positive relationship between the Paris clinical school and the basic sciences than is sometimes indicated by historians of medicine. Erwin Ackerknecht, for example, tends to see "hospital medicine" and "laboratory medicine" as mutually exclusive, and the representatives of the latter as outsiders in the medical world of France in the first half of the century.[2] Such a view is supported by the fact that many medical men prominent in the laboratory sciences were excluded from chairs in the medical faculty, and by the fact that a vocal segment of the medical community held a negative opinion of the role of the sciences in medicine. Magendie complained often enough of these prejudices, and as late as 1865 they were a principal object of Claude Bernard's criticism in his *Introduction à l'étude de la médecine expérimentale*. Yet beginning in the 1820s, Magendie himself was a hospital physician, who fully shared his colleagues' commitment to exact clinical observation and autopsy.

His physiology and studies of drug action were deeply marked by the localism and anatomical concreteness that characterized the pathological anatomy increasingly favored by the clinicians. His interest was overwhelmingly in human physiology, and comparative studies served only as a means to that end. In his program in pathological physiology the clinic, the autopsy room, and the laboratory were joined in a reciprocity of interest. For the execution of this program he had the sympathetic cooperation and support of a substantial number of Paris physicians. Clinical experience and physiology—far from being in simple opposition—found themselves in a relation of mutual benefit that was to carry over into the 1850s and the early years of the Paris Société de Biologie.

The beginnings of Magendie's interest in the nervous system coincided with his earliest experimental studies. His research on the *strychnos* poisons localized their site of physiological action in the spinal cord, and tended to bear out Legallois's fundamental findings on the cord's role as seat of motion. Although much of his research attention in the 1810s focused on absorption and nutrition studies, his investigations of drug action often raised the question of the involvement of the nervous system. There is some evidence to indicate that drug action studies played a significant role in the shift of his primary interest to the nervous system from about 1821.[3]

That shift was undoubtedly eased by the compatibility between the special demands imposed by experimental study of nerves, spinal cord, and brain, and the character of Magendie's experimental physiology as it had evolved by 1821. Sectionings, ablations, and precision dissections of living or dead tissues called for the surgical aptitudes and skills that Magendie presupposed in the physiologist, and that he possessed to an exceptional degree.[4]

Continued access to the veterinary school at Alfort assured a locale for experiments and a supply of experimental animals. The Alfort school and one of its professors, Jean Dupuy, were often mentioned by Magendie in connection with his research in the 1820s.[5]

Another source for Magendie's increasing interest in the nervous system was the comparative anatomy of Georges Cuvier. Among the leading powers of the Academy, Cuvier had from the beginning taken an active interest in Magendie's experimental physiology. He had encouraged Magendie to give his physiology a comparative dimension, and sometimes supplied the younger man with unusual material for experiment or dissection, including on one occasion a crocodile. By the 1810s Cuvier was firmly convinced of the primary functional importance and classificatory value of the nervous system.[6] He had given strong encouragement to Legallois's studies of the spinal cord, and had also reinforced the young Pierre Flourens's first experimental studies of nervous phenomena.[7] Magendie, elected to membership in the Academy in 1821, must have been acutely aware of the Permanent Secretary's sense of priorities.

Magendie's turn to study of the nervous system also coincided with the beginnings of French scientific and popular interest in the work of Gall and Spurzheim. German in origin and medically trained in Vienna, the two arrived in Paris in 1807. Between 1810 and 1819 they published a massive illustrated work on the anatomy and physiology of the nervous system and brain.[8] Their doctrines were novel in several respects. Against the sensationalists, they held that the moral and intellectual faculties were innate. They asserted that these faculties depended on bodily organization, and specifically on that of the brain. Against the Cartesian tradition of the essential unity of the mind, they viewed the brain as a composite of particular organs corresponding to as many particular faculties. Finally, they held that the size and shape of the cerebral organs directly affected the surface conformation of the skull, so that a careful reading of the latter yielded information on the former and thus on a given individual's innate mental faculties. Associated with these doctrines, though by no means forming the basis of them, were anatomical studies of the nervous system represented in a series of excellent plates.[9] By the early 1820s Paris scientific and medical circles were seriously considering the work of Gall and Spurzheim, though commentators often took exception to parts of their doctrine. On at least one occasion commissioners of the Academy of Science, including Cuvier, Duméril, and Pinel, made highly positive allusions to the anatomical parts of that work.[10]

There are numerous references to Gall and Spurzheim in the volumes of Magendie's *Journal,* and the joint work by Magendie and Antoine Desmoulins of 1824 on the anatomy of the nervous system of vertebrates reads in some places like a running com-

mentary on Gall and Spurzheim's treatise of 1810–1819.[11] There
was even some personal contact, for in 1821 Magendie and Spurzheim were both present at experiments performed at Alfort, and
in 1823 both attended the autopsy of an infant who died at the
Hospice des Enfants-Trouvés.[12]

   In the 1820s fundamental differences in methods and conclusions became evident between Magendie on the one hand and
Gall and Spurzheim on the other. Whereas Gall and Spurzheim
distrusted experiment and tended to rely primarily on correlations
between behavior and normal anatomy in humans and animals,
Magendie came to rely primarily on animal experiment and pathological anatomy. Largely through his studies of the cerebrospinal
fluid, Magendie was to become a severe critic of the doctrine of
cranioscopy. Yet there can be little doubt that by drawing scientific
and medical attention to the nervous system, by emphasizing comparative and anatomical methods in its investigation and by ad-
vancing doctrines that might serve as points of departure or as
foils for other researchers, the founders of phrenology played a
role in determining the direction and content of Magendie's early
work on nervous physiology.[13]

   Other individuals played similar roles as objects of emulation,
criticism, or rivalry. In September 1821 Charles Bell's assistant
John Shaw visited Paris to communicate to Magendie some of the
British surgeon-anatomist's views on the nerves. This visit started
the series of events that would lead to Magendie's demonstration
of the sensory and motor functions of the posterior and anterior
spinal roots, and his subsequent priority disputes with Bell.

   In the same year the physiologist Pierre Flourens began a series
of reports to the Academy of Science on his experimental research
on the nervous system, studies in which he was strongly encouraged by Cuvier and which would eventually lead him to important
conclusions on the functions of the cerebellum. From this time
on, Magendie's rivalry with the younger Flourens was to be a spur
to his own studies on the physiology of the nerves.[14]

   In the following year Magendie published a new edition of
Bichat's *Recherches physiologiques sur la vie et la mort,* enlarged with
his own critical notes. This task provided an occasion for a review
of Bichat's experimental investigations on the dependence of other
major organs on the brain, and so may have stimulated or reinforced Magendie's growing interests in this area. His notes to this

edition show that he still had firmly in mind Legallois's results on the dependence of a range of functions on the spinal cord.[15]

The resources that clinical observation and pathological anatomy could place at the disposal of a scientific study of the nervous system were increasingly available to Magendie after 1818, as he gained better access to Paris hospitals.[16] Before his first formal appointment in 1826, Magendie needed the sympathetic cooperation of other Paris physicians to gain entry to clinics and autopsy rooms. That this cooperation was forthcoming in ample measure in the 1820s is evident from numerous references in Magendie's *Journal,* both in his own articles and in the contributions of colleagues in the medical community.[17]

By 1821, then, a number of factors were converging to direct Magendie's interest to the nervous system. The effects were already evident in the first volume of his *Journal.* An article by Scipion Pinel associated a distinct set of symptoms with alteration of the medullary substance of the spinal cord, based on observation and autopsy of two clinical cases. Magendie reported the results of experiments—perhaps undertaken on the occasion of his preparation of the new edition of Bichat's *Recherches physiologiques*—showing that the spinal cord swelled on expiration and subsided on inspiration. And at the beginning of an article summarizing the recent work of Charles Bell, Magendie remarked portentously that "all the facts that throw some light on the functions of the nervous system merit a particular attention."[18]

The form Magendie intended to give this line of research became much clearer the following year. In the second volume of the *Journal* he reported the case of a man in his care who was afflicted by involuntary movements of locomotion, symptoms that yielded to regular small doses of quinine sulphate. The case prompted more general comments. As long as the nervous functions remain intact we suppose them to be of limited number and probably the effects of a single cause, Magendie pointed out. He noted that the phenomena induced by disorders belied that conviction of unity and suggested a multiplicity of nervous actions localized in as many parts of the nervous system. The number and site of these nervous actions and their effects on the other phenomena of life were questions that had been posed but not answered. Magendie conceded that the normal anatomy of the brain, animal experiments, and the effects of pathological disturbance on the structure or functions of the nervous system had all been brought to bear, with varying degrees of success. Of the three approaches, the one based on study of pathological disturbance had, in his view, so far proved most informative, showing for example that lesions of one side of the brain affected the opposite side of the body, and that either sensation or movement might be impaired while the other remained intact.[19]

An article by Pinel in the same volume included remarks giving still more emphasis to the study of pathology as the most promising route to knowledge of normal nervous function. The substance of the article described clinical cases, with autopsies, involving hardening of the nervous system. He had performed no experiments, yet Pinel assured his readers that results like his were of the utmost importance for physiology, since lesions showed that all the moral and intellectual faculties depended on the healthy organization. Loss of organization entailed loss of function. If anything, Pinel went beyond Magendie in his claims for this approach: "The application of pathological physiology to the history of cerebral functions is the only thing that can give positive results; it alone must complete their knowledge; it can correct several errors. This research must therefore merit all our care and attention."[20] In later articles Pinel would speak of "experimental pathology" and would emphasize the role played by experimental physiology in the understanding of nervous disorders.[21]

Read in conjunction with Magendie's remarks, Pinel's statement gave substance and specificity to the title of Magendie's *Journal de physiologie expérimentale et pathologique.* "Pathological physiology" was not to mean, or primarily to mean, the experimental study of pathological processes. Rather it meant, for Magendie and Pinel, the elucidation of normal functions through study of the behavioral and anatomical changes associated with their disorders. As such, it was a concept standing midway between Magendie's experimental physiology and the pathological anatomy just then being developed by the Paris clinical school.[22] In effect, it was pathological anatomy placed in the service of physiology. For Magendie and Pinel it was to be the principal key to the study of the nervous system.

Scattered references in Magendie's publications of the 1820s reveal other aspects of his general program. Ideology, as the study of consciousness, was to be separated from physiology, conceived as the anatomical, experimental, and pathological study of the nervous system in its effects on behavior. In Magendie's view, nothing was known about the way in which intellectual phenomena were produced in the brain.[23] Similarly, good physiology required a clear distinction between what was physical and what was purely vital in the organism. Although Magendie believed in pushing physical explanation as far as possible in physiology, he also left the door open for the existence of irreducibly vital phenomena. Eventually the nervous system would be placed in that category.[24] Consistent with his long-standing views on the contemporary state of physiology, he believed that it was premature to attempt an integrated picture: "Whoever would systematize the nervous functions today would build on sand."[25]

As might be expected in view of Cuvier's special interest in the nervous system, the anatomical part of Magendie's program was given a comparative dimension from the beginning. Though this was not explicit in Magendie's programmatic statement of 1822, it was reflected in articles of comparative interest regularly printed in the *Journal*. The same volume for 1822 included a joint article by Magendie and Desmoulins on the anatomy of the lamprey, and a memoir by Desmoulins on the nervous systems of fishes. That year Desmoulin's memoir had shared the Academy of Science's prize in experimental physiology, with Magendie serving as a member of the prize committee.[26]

Magendie apparently saw in the young physician's interest and aptitudes a means to the discharge of his responsibilities toward comparative anatomy, studies for which he lacked time, enthusiasm, and special ability. Whatever the reasons, the two men became collaborators in the period 1822–1825, and it was Desmoulins who gave the clearest and strongest statement on the importance of comparative studies for physiology. In a memoir of 1824 on the lacrymal apparatus and nervous systems of snakes, Desmoulins described three methods of studying physiology—namely, experiment, pathological anatomy, and study of the different combinations of organs found in the animal series as a whole. He claimed to be the first to follow the third method, though it had been set forth by an "illustrious zoologist"—an obvious reference to Cuvier—about twenty years earlier. He subsequently explained that although at first he tried to find in comparative anatomy only justifications for applying the results of animal experiments to man, he had soon found it to be a source of physiological knowledge distinct from and complementary to animal experiment and pathological anatomy. This was henceforth to be Magendie's formal position.[27]

Cuvier was clearly the major motivating force behind this dimension of Magendie's program, and both Magendie and Desmoulins sided with Cuvier against Geoffroy Saint-Hilaire in the debates over unity of type.[28] Yet they could also be critical, as shown in their dissent from the majority judgment of the Academy of Science on Flourens's early work.[29] Magendie had serious reservations regarding what could be learned about human physiology from study of the animal kingdom below the level of the higher mammals.[30] Even more fundamental was the divergence of ultimate goals that such reservations revealed. Whereas Cuvier encouraged a comparative physiology that would embrace the whole animal series and that was thus essentially biological, Magendie saw comparative studies first of all as a means to the understanding of human physiology, itself conceived as the scientific basis of medicine. It was an orientation that was to persist for the remainder of his career.

Since Magendie lacked an official teaching position in the 1820s, the principal vehicles for the communication and promotion of his research were his private courses in experimental physiology and, above all, the *Journal.* The content of the latter reflects both the range of approaches and subject matters that he thought could be brought to bear on physiology, and the changing direction of his own leading interests. Not only experimental, clinical, pathological, and comparative studies but also chemical, physical, and embryological research found a place in its pages. Magendie's growing interest in the nervous system was reflected in the disproportionate number of articles on the subject by himself and others in the volumes for 1822 and 1823.[31]

Formally, the program for study of the nervous system that emerged in Magendie's writings of the 1820s was tripartite, based on coordinated experimental, clinical or pathological, and comparative methods. In fact, it was already clear that the experimental and the clinical or pathological approaches would dominate. The nature of Magendie's physiology as it had developed prior to 1821, his continuing medical practice and increasing access to hospitals, the emphasis he gave to the concept of pathological physiology, his view of comparative studies as a means to the end of human physiology, and his delegation of these studies to a younger colleague all point in this direction. The leading position of experiment is nowhere better displayed than in the work that led Magendie to the most important single discovery of his career.

In 1822 Magendie published descriptions of experiments supporting the conclusion that the posterior (dorsal) roots of the nerves proceeding from the spinal cord are primarily associated with sensation, whereas the anterior (ventral) roots of the same nerves are primarily associated with motion.[32] Now regarded as a classic, this paper touched off controversies in scientific and medical circles that carried over into the twentieth century. No one doubted the importance of Magendie's finding, which immediately opened new paths to the functional analysis of the nervous system. The issue was priority. The British surgeon Charles Bell—or those speaking on his behalf—claimed that Bell had made the discovery as early as 1809. The issue has long since been settled in Magendie's favor, though its memory survives vestigially in the eponym "Bell-Magendie law."[33] What is of interest here is just what has been lost sight of or passed over in other commentaries: the means by which Magendie achieved and demonstrated his conclusions. In this instance the surgical qualities of his physiology were crucial to its success. Clinical or pathological experience played a complementary but secondary role, comparative thinking none at all.

As was often the case with problems taken up by Magendie, prior work on the function of the spinal cord was suggestive but inconclusive. The distinction of motor and sensory nerves was at least as old as the Alexandrian Herophilus (c. 300 B.C.), but had never been investigated in detail. Galen (A.D. 200) performed experiments on the spinal cord, but had no successors. Experimental study of spinal cord function was revived only in the eighteenth century, in the work of Robert Whytt and others. Working in the first decade of the nineteenth century, Magendie's contemporary Legallois showed that the life of any part of the body depended on the integrity of the part of the spinal cord giving rise to nerves supplying that part. Legallois spoke of the cord's control of sensation and voluntary movement, but did not assign these to anatomically distinct locales.[34]

Meanwhile, across the Channel, Charles Bell had also begun work on the anatomy and physiology of the nervous system. By

1810 he had sketched the outlines of a scheme based on the idea that the cerebrum and cerebellum are centers of different functions. In Bell's view the cerebrum was the seat of sensation and motion, the cerebellum the seat of "secret," vital, or involuntary actions. He saw, in the apparent extension of the cerebrum into the anterior part of the spinal cord and anterior roots and of the cerebellum into the posterior part of the cord and posterior roots, evidence of their divergent functions. He found further evidence in experiments on the spinal roots that he performed as early as 1810. Opening the spine of an animal, he "pinched and injured" the posterior roots of the nerves and observed no muscular motion. When he touched the anterior roots the associated parts were immediately convulsed. He obtained the same results when he injured the posterior and anterior parts of the spinal marrow. He concluded that the anterior roots were sensitive, the posterior insensitive—nearly the opposite of what Magendie was later to demonstrate. Bell subsequently revealed that because he had not wanted to inflict unnecessary pain, he had experimented on stunned rabbits. Thus, he could not have observed sensitivity in the usual sense in either posterior or anterior roots.[35]

Bell expected great things of his ideas on the functional divisions of the brain and the associated experiments on the spinal roots. On several occasions he compared himself to Harvey.[36] He made little effort to publicize his ideas, however, and did not extend his experimental results.[37] He subsequently concentrated on the facial and trigeminal cranial nerves. In 1821 he read to the Royal Society a paper in which he discussed the fifth and seventh cranial nerves as possible respiratory nerves within his system of classification, concluding that the seventh cranial nerve belonged in this category.[38] It was at this juncture of Bell's work that his assistant and brother-in-law John Shaw visited Paris, in September 1821.[39] Shaw explained the essentials of Bell's ideas on the nervous system to Magendie, and repeated his experiments on the facial nerves at Alfort in the presence of Magendie, Spurzheim, and the veterinarian Jean Dupuy.[40]

Nine months later, in June 1822, Magendie published his first paper on the functions of the spinal roots.[41] He was vague about the source of his interest—he had wanted to try the experiment of cutting the posterior roots "for a long time," he said—and claimed to have approached its results without preconception.

There can be little doubt, however, that in the framing of the problem and in the timing of his experiments Magendie was indebted to Bell. After Shaw's visit, Magendie had published a summary of Bell's ideas which stressed that every regular nerve has two roots, one anterior and one posterior, and that these have different functions, associated with the cerebrum and cerebellum, respectively.[42] Shaw had given Magendie a copy of his laboratory manual, which included a note indicating a difference in the degree of sensibility between the posterior and anterior spinal roots.[43] Magendie was favorably disposed to Bell's work.[44] Such assertions and hints might well have led him to try his own hand at the spinal roots, though at that time he still had no detailed knowledge of Bell's experiments, and the outcome of his own investigation was to be very different from that of Bell's. Magendie later conceded that Bell, "led by his ingenious ideas on the nervous system," was very close to discovering the functions of the spinal roots.[45]

Whatever prompted Magendie's experiments, their execution and the conclusions he drew from them were entirely his own. The challenge, as Magendie presented it, was essentially surgical.[46] How could the posterior spinal roots be sectioned without injuring the cord? And how could the anterior roots be sectioned without injuring the intervening posterior ones? After a series of failures, Magendie succeeded in reaching his first objective. Operating on small young dogs, he was able to lay bare the membranes of the posterior half of the spinal cord with a single stroke of a very sharp scalpel. Dividing the dura mater, he exposed the posterior roots of the lumbar and sacral pairs. Raising the posterior roots on one side successively with the blades of a small scissors, he could section them without harming the cord. Closing the wound with a suture and observing the animal, he found that the operated side had become completely insensible, while its motor function was preserved.

Magendie's second task at first seemed impossible. How could he expose the more deeply embedded anterior part of the cord without somehow involving the posterior roots? So anxious was he to make the comparison, he said, that "I did not stop dreaming of it for two days, and finally I decided to try to pass in front of the posterior roots a type of cataract knife, the very narrow blade of which permitted cutting the [anterior] roots by pressing them with the edge of the instrument on the posterior face of the body of the vertebrae." This maneuver did not work, because it opened large veins on the side of the canal. But "in making these attempts," Magendie went on, "I perceived that in pulling on the vertebral dura mater, I could catch a glimpse of the anterior roots joined in fascicles, just where they penetrate this membrane. This was all I needed, and in a few moments I had cut all the pairs that I wished to divide."[47] In this case, the member on the operated side was completely immobile and flaccid, but retained an "unequivocal sensitivity."

Magendie further established that section of both anterior and posterior roots produced an absolute loss of sensation and movement on the operated side. He repeated and varied the experiments on several species of animals, always obtaining the same results. On this basis he was prepared to assert definitely the association of sensitivity with the posterior roots, and of movement with the anterior ones.

Surgical ability, instruments, and imagination played a decisive role in this discovery, and the same qualities were manifest in Magendie's attempts over the next year to extend his findings. The most important addition was his recognition that some movement could result from stimulation of the posterior roots, some sensation from stimulation of the anterior roots. Thus, the division of function he had established, though marked, was not absolute.[48] Pursuing the nerve roots to their sources in the spinal cord, Magendie found that the anterior and posterior sides of the cord had the same properties as their respective roots. Reflecting on the possible extension of these properties into the brain, he had occasion to criticize both the method and the conclusions of Charles Bell, though the British surgeon was not cited by name. Anatomy alone, Magendie noted, indicates that the anterior roots and the anterior part of the cord connect to the cerebrum, and therefore that the cerebrum should be the seat of motion. The posterior part of the cord and the posterior roots are joined to the cerebellum, and by this reasoning the latter should be the seat of sensation. "But anatomy is not enough," he admonished, "physiology [animal experiment] and pathological facts must confirm the indication; and up to now neither of these means has established what anatomy appears to show in such an obvious way." Lesions of the cerebellum did not bring loss of sensitivity, nor did removal of the cerebral hemispheres necessarily entail loss of movement.[49]

An instance of the confirmation that Magendie sought in pathological facts was provided by Antoine Royer-Collard, a professor at the Faculty of Medicine and physician to the *maison de santé* of Charenton.[50] Royer-Collard described the case of a male patient who had entered the Charenton in 1806, and who had died there in 1823. The man had suffered from a progressive paralysis of the limbs that left their sensibility intact. Autopsy revealed extensive pathological alteration of the anterior part of the spinal cord and anterior nerve roots. The posterior surface of the cord and its membranes were in healthy condition. Magendie drew the obvious conclusion in favor of the distinction he had defined.[51]

In the course of his experiments on the spinal roots, Magendie had sometimes noticed a flow of liquid from the cord when the surrounding membranes were opened.[52] Between 1821 and 1825 this observation converged with others to become part of Magendie's second major research achievement of the 1820s: the discovery of the cerebrospinal fluid.

In 1825 Magendie presented evidence that a "transparent, serous fluid" is normally present in the spaces that separate the spinal cord and brain from the membranes that surround them. Others had occasionally noted the presence of a fluid in these locales, but in the conventional anatomy of the day it was considered to be a product of pathological processes.[53] Magendie emphasized that it was only in the course of animal experiments that he had been able to confirm the normal presence of the fluid.[54] In fact, it appears that his first awareness of the fluid and subsequent confirmation of its normal presence developed through a close interaction between experimental and pathological findings.

As early as 1821 Magendie studied movements of the spinal cord isochronous with respiration, and in a note on the same subject added to the 1822 edition of Bichat's *Recherches physiologiques* he remarked that the spinal cord is an organ placed "in the midst of a cavity that it does not fill completely."[55] The same edition of Bichat included more than one reference by Magendie to Legallois's experimental work on the spinal cord, and the *Journal* for 1822 printed his first paper on the various functions of the spinal roots.[56] It is clear that by 1823 Magendie's research was focused not merely on the nervous system but specifically on the spinal cord and brain.[57]

A pathological case reported by the physician P. Rullier in the

Journal for that year reinforced the same tendency. Rullier's patient had suffered from unusual symptoms, including progressive
rigidity and loss of the use of the arms, a growing tumor and
deformation of the spine, and pain in the upper limbs and trunk.
Unable to diagnose, still less to arrest the course of the disease,
the attending physicians could only try to make the patient as
comfortable as possible. When the illness proved fatal, a careful
autopsy was performed. The most striking result was the finding
that the upper part of the spinal cord, in the lower cervical and
upper lumbar regions, was softened to such an extent that it appeared that the dura mater of this area was filled with fluid. A
small opening made in the dura mater allowed a large quantity of
fluid to flow. Magendie, who attended the autopsy, noted that this
liquid was almost colorless and that all those present thought it a
result of a hydropsy of the cord itself. He referred to the fluid as
"serosity." His primary interest was not in the fluid, however, but
in how, given its apparent replacement of the substance of the
cord, nervous connection was maintained between the brain and
the lower part of the cord and its associated nerves. Yet he saw
the case as still another indication of contemporary ignorance of
the physiology of the nervous system, and it served to draw his
attention once more to the spinal cord and its associated phenomena.[58]

In 1823 Magendie also attended the autopsy of an infant who
had died at the Hospice des Enfants-Trouvés. Gilbert Breschet,
chief of anatomy at the Faculty of Medicine, was in charge; Johann
Spurzheim and a Dr. Nicati were also present. The autopsy revealed, among other things, about twenty ounces of "serosity" in
the skull, and only traces of a brain. Serosity was also present in
the spinal cavity, but was found to be more albuminous than that
in the skull. Moreover, the spinal cord occupied its ordinary volume. Breschet ventured the general remark that in subjects in
which the brain and spinal cord do not form or form only incom-
pletely, they are replaced by a "more or less albuminous serosity"
which fills the brain and thus favors the development of the bones
in this part of the head. This serosity, he thought, was found in
the cavity of the arachnoid—that is, the serous pocket located
between the dura mater and the pia mater.[59]

Only in the following year, 1824, did Magendie turn his attention directly to the fluid associated with the brain and spinal cord,

and recognize and confirm its normal presence. His private course on experimental physiology for that year emphasized the functions of the nervous system.[60] Desmoulins later stated that it was in the middle of 1824, while incising the dura mater above the fourth ventricle in living quadrupeds, that Magendie discovered that the cavity of the dura mater was filled with water.[61] And Magendie's *Journal* for that year printed a report of a clinical case with autopsy by a Dr. Koreff in which the pathological condition was associated with a large quantity of fluid in the upper part of the spinal cord. Although he saw the extraordinary quantity and gelatinous consistency of the fluid as products of chronic inflammation, Koreff was prepared to interpret these phenomena as simply the very much altered properties of the fluid that Magendie had found to be "constantly" present in the vertebral columns of living animals and of humans killed by accident and immediately opened.[62]

This series of events is remarkable for the close interaction and mutual reinforcement of experimental, clinical, and pathological findings, to the virtual exclusion of comparative studies. Desmoulins did publish a note in 1824 on the fluid contents of the lamprey spinal cord. The following year he made reference to his discovery of 1821 that the spinal cords of fishes were protected from external shocks by a surrounding layer of fluid. Yet it appears that no one, including Magendie, was affected by the earlier findings, and the later note was concurrent with and perhaps prompted by Magendie's own research on the subject. In the 1825 monograph on the nervous systems of vertebrates, Desmoulins treated the cerebrospinal fluid as something already known.[63]

The same almost exclusive reliance on pathological and experimental approaches appeared in Magendie's first memoirs on the cerebrospinal fluid, published in 1825 and 1827. In the first of these he noted that he had observed a transparent serous fluid between the spinal cord and its surrounding membrane on several occasions, but it was only when he had turned to animal experiments that he had been able to confirm its normal presence. A small opening made in the exposed vertebral dura mater of any mammal invariably released a certain quantity of the fluid, and similar results could be obtained by opening the membrane over the brain or cerebellum. Such experiments were necessary because the fluid tended to recede after death. Since police regulations in Paris forbade autopsy in the first twenty-four hours after death, the opening of cadavers would not necessarily confirm the presence of the fluid.[64]

Once he had established its presence in living mammals, however, Magendie was anxious to extend his finding to humans. Taking advantage of a temporary assignment to a medical service of the Hôtel-Dieu late in 1824, he carried out a series of autopsies with this end in mind. Bypassing the standard autopsy procedure in which the brain was opened and dissected first—a method that would allow any fluid to escape—he began instead with the spinal column and exposed it along its entire length, leaving its dura mater intact. In every case he found the membranes surrounding the cord to enclose a transparent liquid, which communicated freely with another liquid covering the surface of the cerebellum and brain. Since this finding was uniform regardless of the cause of death, Magendie concluded that contrary to the prevailing opinion of contemporary medicine, it represented a normal condition. Further careful study of human cadavers revealed that the fluid was not located in the cavity of the arachnoid membrane but rather in the space between the inner layer of the arachnoid and the pia mater. This finding again violated prior expectations, since contemporary anatomy, following Bichat, regarded the arachnoid as the serous membrane of the brain and spinal cord.[65]

By his own testimony, Magendie devoted most of his efforts in the next two years to further study of the cerebrospinal fluid. These investigations were greatly facilitated by his appointment in 1826 as substitute physician at the Salpêtrière, a position he was to hold until 1830. A regular hospital post provided ready access to a continuous stream of clinical and pathological material, and although the memoir of 1827 included experimental as well as pathological arguments, it is not surprising that Magendie presented his findings as having reference mainly to human physiology.[66]

Magendie was now prepared to move beyond a bare assertion of the normal existence and locale of the cerebrospinal fluid to try to assess its production and role in the animal economy. The fluid, he thought, was the means by which a stable pressure was maintained between the brain and the skull, whereas in the abdomen the walls themselves varied in volume. When he removed the fluid from the spinal cavity of living animals he found that it was quickly renewed, and the site of its renewal was the pia mater, not the arachnoid. Removal of the fluid in dogs produced marked behavioral disturbances, as did an abnormal increase in its volume or pressure. Magendie noted that the latter changes, induced by experiment, were reproduced pathologically in human victims of the disease spina bifida, in which a collection of fluid on a part of the spine caused temporary increases of pressure on the cord. Referring again to pathological material, he suggested that it was the pressure of the fluid that maintained the size of the skull and bony cavities, especially in the absence of parts of the brain or cord.[67]

Magendie held that in addition to its role in the maintenance of pressure, the fluid had other physical properties that merited attention. He explored the effects of temperature change by animal experiments in which he first withdrew the fluid, then reinjected it at a lower temperature. Magendie persuaded the chemist Chevreul to conduct a chemical study of the fluid for humans, and Jean Lassaigne was to undertake the same study for horses. Magendie was convinced that the extensive contact of the fluid with the central nervous system must be associated with important electrical phenomena, in the study of which Antoine Becquerel had agreed to assist him.[68]

What was the relationship between the cerebrospinal fluid and the liquid often found in the ventricles of the brain? In particular, was there an anatomical communication joining the ventricles and the fluid of the spine? The lower part of the fourth ventricle was a likely location for such an opening, but Bichat's description of the area ruled this out. Once again, clinical and pathological experience provided the answer. Magendie recalled an autopsy performed on one of his wards during an assignment at the Hôtel-Dieu, a case that showed a clear communication between the fourth ventricle and the cerebrospinal fluid. The finding was confirmed by a series of autopsies of normal brains. Moreover, in more than forty autopsies performed following disease in which there was more or less dilation of the ventricles, there was invariably direct communication between the liquid of the brain and the spine. Experiments on cadavers supported the conclusion that such communication was normal.[69]

Since contemporary anatomy did not recognize the normal presence of fluid in the brain, Magendie carried out more than fifty autopsies during 1826 to verify this point alone. It was the excess of the fluid, he argued, that led to disorders. The probable source of the fluid in the pia mater of the spinal cavity was suggested and confirmed by a mutually reinforcing series of clinical, pathological, and experimental findings.[70]

The medical milieu, including the sympathetic cooperation of several other physicians (he referred in passing to the successful research of his colleague Charles Ollivier on the spinal cord), had played a fundamental role in these investigations. Magendie was not confident that his results would be used by the medical community, however, though knowledge of the cerebrospinal fluid was bound to throw light on the most serious and frequent diseases of the brain, such as apoplexy and cerebral fevers. He saw that medical empiricism, a prior commitment to systems, or simple disbelief in the practical relevance of anatomy and physiology would inhibit its assimilation.[71]

For those physicians ready to accept his results, a vista was opened for further research on the way in which the fluid was renewed, on its fate in humans at various ages from the embryo to death, on its variations in quantity and quality in health and disease, on its possible influence on the intellectual faculties, and on its existence and role in the series of vertebrates. A cooperative endeavor was needed, an endeavor that Magendie clearly expected to lead but to which he was also inviting physicians, chemists, veterinarians, and comparative anatomists.[72]

Magendie's preoccupations of the moment were clearly reflected in his annotated edition of Bichat's *Traité des membranes,* also published in 1827. "Brilliant, though flawed" when it first appeared, Magendie admitted, the book was now a source of serious error on some points, if left uncorrected. Magendie was only too happy to perform this task, and it is not surprising to find that Bichat's most egregious error was a failure to recognize the cerebrospinal fluid. It was a failure that affected simultaneously his anatomy, his physiology, and his pathology. He did not know of the opening from the fourth ventricle to the subarachnoid space of the spinal column. He described as hydropsy of the arachnoid what was simply the cerebrospinal fluid in normal or excessive quantity. He did not realize that in any case the fluid is not contained in the arachnoid cavity but between the free inner layer of the arachnoid and the pia mater of the cord. He did not realize that accumulations of fluid were in fact rare in the arachnoid cavity.[73]

By the time he read his "Mémoire physiologique sur le cerveau" to the public meeting of the Academy of Science the following June, Magendie was still more confident and expansive. Though it was largely a recapitulation for a general audience of earlier findings on what Magendie called, with proprietary affection, "my liquid," the memoir did contain some new substance. Most important, it introduced Magendie's ill-fated suggestion that the pineal gland functioned as a mechanical stopper controlling the flux and reflux of cerebrospinal fluid between spinal cord and brain. Magendie held that Gall and Spurzheim's elaborate system of cranioscopy, based on a presumed correlation between surface conformations of the skull and brain, was refuted by the bare fact of varying thickness of the fluid throughout the various regions of the brain and spinal cord.[74]

Though they are nowhere mentioned by name, the thinking of the same authors may have played a role in Magendie's formulation of and tentative answer to a new question—namely, the influence of the cerebrospinal fluid on the faculties of intelligence. As a first approach, Magendie had attempted to determine the relative quantities of the fluid in people in possession of their reason, in imbeciles, and in the insane. The needed information came from autopsies performed at the Salpêtrière. Both idiots and the insane were found to have greater than normal quantities of the fluid, but differently disposed. A genius who died in old age had less. These results suggested the general conclusion that the development of mental faculties was in inverse proportion to the quantity of cerebrospinal fluid.[75]

The 1828 memoir marked the end of the first phase of Magendie's research on the cerebrospinal fluid. He had recognized its normal presence, determined with precision its location and site of production, and had begun to arrive at tentative conclusions regarding its physiological role. All of this had been achieved through an interplay of clinical experience, pathological observation, and animal experiment. For the first two of these, Magendie's increasing access to Paris hospitals and the cooperation of physicians sympathetic to his program of pathological physiology proved decisive. For the experimental component, the Academy of Science provided a stimulus, Magendie's private courses in physiology supplied on occasion, and the veterinary school at Alfort frequently made available a locale and experimental material. For all aspects of the work, the *Journal* offered a vehicle.

Whatever the role of Cuvier and comparative anatomical studies in the initiation of Magendie's work on the nervous system, their part in the process of discovery and confirmation had proved negligible. Desmoulins's comments on the cerebrospinal fluid in fishes stood in lonely isolation. In contrast to his obvious excitement with present clinical and pathological findings, Magendie's references to comparative studies seem almost an afterthought, perhaps included out of deference to his old patron, the director of the Muséum.

Discovery of the cerebrospinal fluid involved not merely the addition of new facts but also a reevaluation and fresh perception of facts already known. This recognition served, at least momentarily, to jar Magendie out of the naive empiricism that he had often expressed in association with earlier work. Facts, he now conceded, were not enough. The facts needed for recognition of the cerebrospinal fluid had been available for many years, but because people were not mentally prepared, they did not see their importance: "The vision of the senses is short; the vision of the mind alone may be great in extent."[76]

When he received a permanent appointment to the Hôtel-Dieu in late 1830 and to the chair of medicine at the Collège de France the following April, Magendie had not lost interest in the nervous system or the cerebrospinal fluid, though by that time those interests were being subordinated to other activities.[77] His duties as physician and teacher became more pressing. For a time, the cholera epidemic absorbed much of his energy in both capacities. Clinical and experimental studies of drug action, embodied in new editions of the *Formulaire,* continued to claim time and attention. By the mid-1830s a special interest in the physics of the circulatory system, conceived as part of a general study of the physical phenomena of life, was taking precedence over other involvements. Jean Poiseuille, whose 1828 thesis on the force of the heart had been printed in Magendie's *Journal,* now became his close collaborator.[78]

It was not until 1838 that Magendie again directed the major part of his attention to the nervous system. Between December of that year and the following July, he devoted his lectures at the Collège de France entirely to presentation of past and present research in this area. The methods he outlined to his students, and their relative importance, did not differ markedly from those that had characterized his work in the 1820s. The dependence of the nervous system on special vital laws would, Magendie said, limit the roles of physics and chemistry. He would attempt to bring anatomy, especially the recent discoveries in microscopical anatomy, into relation with nervous functions, but details of nervous structure—which in Magendie's view had yielded little information on function—would not be objects of study in themselves.[79]

By the time he gave these lectures, Magendie had become familiar with much of recent German research and was able to make favorable mention of the work of Gabriel Valentin, Jan Purkinje, Gottfried Treviranus, Karl Burdach, Robert Remak, and others, especially in microscopy. He thought highly of Christian Ehrenberg's studies of infusoria, though he chided the German for attributing to these creatures too high a degree of organization. It was otherwise with nature philosophers such as Lorenz Oken and Julius Carus, who had produced "bizarre conceptions" and nothing but pure speculation on the nervous system. Similarly he rejected Gall and Spurzheim's ideas on the organization of the brain and on cranioscopy, though he was prepared to acknowledge that the principal cerebral functions might one day be localized in the central nervous system.[80]

Magendie admitted that comparative studies might be a means to the knowledge of the functions of certain nerves. Yet his reservations about the value of such studies for human physiology were at least as strong as they had been two decades earlier. He pointed out that it was not yet determined that the invertebrates possessed a nervous system. Even for the vertebrates, knowledge of which parts corresponded between one class and another was often uncertain, and conditions of life could and did differ radically. Only vertebrates closest to man in organization and mode of existence could be relied on for such investigations.[81] No more than in the 1820s did Magendie conceive of a general science of biology.

Whatever might be said of other methods, it is clear that Magendie's program was still dominated by the notion of pathological physiology. The dual physiological and pathological interest was reflected in the title of the published lectures. In the lecture that opened the course, he expressed the hope that his results would be of use at the patient's bedside, a sentiment which for Magendie always represented urgent practical concern rather than pious wish. He expected positive results to emerge primarily from coordination of clinical and pathological observation and animal experiment. Since experiments on human subjects were impossible, pathology was crucial to the understanding of specifically human nervous function. Properly studied and analyzed, clinical, pathological, and experimental findings could not be contradictory.[82]

In these respects, the lectures on the cerebrospinal fluid were representative. The subject was given disproportionate attention, in part because Magendie perceived that his discoveries had not yet been generally recognized or assimilated by the medical and scientific community. He attributed this in part to the "barbarous" methods used to expose the brain and spinal cord in standard autopsy procedure, in which the violent blows of a steel hammer often broke the enveloping membranes and released the fluid prematurely. He suggested equally simple but gentler dissection techniques that would leave the outer membranes intact, and described a new method of autopsy to be used when the goal was specifically the collection and measurement of the cerebrospinal fluid. Because the quantity of fluid present decreased with increase in the interval between time of death and autopsy, different anatomists had obtained different results. This was a second reason for lack of recognition of its normal presence. Finally, Bichat's doctrine on membranes, according to which the arachnoid was a serous membrane, had blocked recognition of the true locale of the fluid between the inner layer of the arachnoid and the pia mater.[83]

Despite the special attention accorded the cerebrospinal fluid, Magendie had little that was fundamentally new to report on its physiology, and most of the experiments he described dated from the 1820s.[84] What was novel was the great range of pathological material either marshaled in support of physiological results established earlier by experiment or presented as phenomena that those results might help explain. Drawing on more than fifteen years of experience on hospital wards and in autopsy rooms, Magendie was now able to give formidable substance to his program of pathological physiology.

Discussing his discovery of the passage between the liquid of the spine and brain at the fourth ventricle, Magendie displayed two plates for his class. One of these, showing an enormous dilation of this opening, was drawn from a patient who had died of hydrocephalus. The other showed the opposite condition, the complete obliteration of the opening in a mentally ill woman. In the latter case, Magendie suspected that the patient's illness had been associated with the abnormal condition of her fluid. More striking was the way in which pathological anatomy was here brought to the support of physiology by providing instances of a normal phenomenon exaggerated in two different directions.[85] In a later lecture, Magendie supported the same point by opening a cadaver in which the cerebrospinal fluid had been allowed to freeze, to reveal its continuity; one instance of his frequent resort to experiments on cadavers.[86]

Once established, the normal existence and properties of the cerebrospinal fluid opened new possibilities for the understanding and diagnosis of diseases affecting the nervous system. Magendie pointed out that it was now necessary to distinguish different forms of hydrocephalus, depending on whether the excess of fluid was located on the ventricles or on the outer surface of the brain. His prediction that blood would be present in the cerebrospinal fluid in a case of apoplexy was confirmed by autopsy. He invoked the existence of fluid pressure on the brain to reconcile an apparent contradiction between the results of experiment and a clinical case involving motor function. In another case, he suggested that compression of the brain may have been caused by the cerebrospinal fluid.[87]

The clinical context of Magendie's work on the nervous system was also reflected in the surgical qualities of his approach. In demanding of the physiological experimenter a manual dexterity and presence of mind due in part to native ability and in part to long practice, he came very close to making the parallel explicit. More important, those qualities were repeatedly embodied in research and demonstration procedures.[88]

Frustrated by what he perceived as the continuing failure of the medical community to assimilate his discoveries on the cerebrospinal fluid, Magendie brought together his findings in *Recherches physiologiques et cliniques sur le liquide céphalo-rachidien ou cérébro-spinal* (1842).[89] The *Recherches* was divided for rhetorical or pedagogical purposes into three parts, two of them corresponding to the physiological and clinical divisions of the title, the third reporting the results of comparative studies. On the whole, the treatment paralleled that of the lectures both in methodological priorities and in substance. If anything, the pathological side was expanded, probably with a medical audience in mind. Magendie listed pathological conditions that had to be reexamined in connection with the cerebrospinal fluid. They included acute or chronic hydrocephalus, accumulation of fluid on the spinal cord (*hydrorachis*), serous apoplexy, the various meningitises, hemorrhages, and softenings of the brain or cord. In his view, this part of clinical medicine had to undergo a complete revolution.[90]

That revolution involved primarily the abandonment of the older and still current view that the bare presence of fluid in the spinal cavity and ventricles of the brain represented a pathological condition. Doctors had to learn to see that, on the contrary, the presence of the fluid in a certain quantity and with certain chemical and physical properties was normal. What was pathological was the augmentation or diminution of this quantity, or the alteration of these properties beyond definite limits. Such changes were often the secondary products of other pathological phenomena, and once induced they could give rise to further disturbances. Excessive quantity of the fluid could be the result of blockage of the passage from ventricles to spinal cavity by tumors, or of impeded venous resorption due to some constraint on the venous system, or simply of the body's compensation for decrease in brain size with age. An abnormally low quantity could be due to unusual growth of all or part of the brain, or to thickening of the bones of the skull, both conditions leading to displacement of fluid. Pathological destruction of a part of the substance of the brain or cord could lead to its replacement with cerebrospinal fluid. Its chemical and physical properties could be significantly modified either by the introduction of foreign substances such as drugs or by pathological changes that might alter the proportions of its constituents or introduce into it solid substances such as blood, fragments of nervous tissue, or pus.[91]

Acceptance of this range of possibilities would require of physicians a new precision in diagnosis and nomenclature, and the abandonment of worthless metaphors. Vague phrases such as "serous apoplexy" would have to be replaced by terminology that reflected exact knowledge of the underlying process, in this case the rapid accumulation of cerebrospinal fluid in the ventricles.[92]

Behind the reluctance of the medical community to commit itself to Magendie's conclusions on the cerebrospinal fluid lay the more fundamental issue of the credibility of experimental methods in physiology and medicine. In a lecture given in January 1839, Magendie referred angrily to a recent debate in the Académie de Médecine in which disagreements among experimenters had been cited as evidence of the irrelevance of experiment for clinical medicine. In words that were later to be echoed by Bernard in the *Introduction à l'étude de le médecine expérimentale* he affirmed his faith in the ultimate consistency of experimental results among themselves and with clinical observation.[93]

The reaffirmation, though genuine, was now more nuanced. For if the controversy within the medical community over the value of animal experiment had any positive role, it was to further dissipate the naiveté of Magendie's early career on the univocality of experimental results. He was now willing to admit that some experiments were more equal than others. A physiological experiment had to be done with the proper technique. This in turn required a manual dexterity and presence of mind that were in part the products of training and practice, in part the expression of natural gifts. Once carried out, the results of an experiment could be properly understood only in the light of past experience. Far from being a relatively easy source of unambiguous facts, experiment was now conceived as a delicate and easily mishandled instrument, "a true art."[94] When he made this argument, Magendie was just beginning research that would lead to its classic exemplification—research that resumed and extended his old work on the spinal nerve roots.

In experimental demonstrations prepared for his class at the Collège de France in 1839, Magendie found that pinching or cutting the anterior spinal nerve roots in dogs produced evidence of pain, and thus of sensitivity. The result was not new, for Magendie had already reported in one of his earlier papers on the spinal roots that the anterior root was not exclusively motor.[95] He now took the problem a step further, by seeking the source of that sensitivity and finding it not, as he expected, in the central connection of the anterior root to the spinal cord but in its peripheral connection to the posterior root. He called the phenomenon "recurrent sensitivity," since the property evidently "circled back" from the periphery to the anterior root after traveling outward from the cord through the posterior root.

The discovery of recurrent sensitivity became a controverted issue in French scientific and medical circles in the 1840s. In part this was due to the physiological significance attached to the phenomenon by Magendie and his young collaborator, Claude Bernard.[96] More important were the contradictory results obtained subsequent to Magendie's 1839 demonstrations. In the early 1840s other investigators, and Magendie himself, were unable to confirm the existence of recurrent sensitivity. These contradictory findings raised anew the issue of the credibility of experimental methods in physiology and medicine. They forced Magendie and Bernard to refine their experimental control of the phenomenon. Ultimately they led Bernard to reflect on physiology as the operative determination of the conditions of phenomena, or what he would later call "experimental determinism."

Magendie established that the posterior root was the source of the sensitivity of the anterior in a series of experiments that, in final form, were elegant in their economy.[97] He sectioned a dog's anterior spinal root in the middle and pinched in turn its spinal and peripheral ends. The former action had no effect on behavior, while the latter produced symptoms of acute pain. In a second experiment, Magendie cut a posterior root in the middle. Again he pinched in turn each of the cut ends. This time the stimulus applied to the spinal end caused pain, while the peripheral end gave no evidence of sensitivity. These and other experiments converged to point to the same conclusion: that the property of sensitivity passes from the cord through the posterior root to the periphery, then returns to the anterior root. Pleased with the coherence of his findings, Magendie reflected that "when we arrive at contradictory results, that is because of the inadvertence or lack of dexterity of the operator, and not the inconsistency or incertitude of the phenomena."[98]

Magendie's confidence on this score was soon to be severely tested, and by one of his own students. Achilles Longet, like Claude Bernard a product of the Paris Faculty of Medicine, aspired to be at once an experimental physiologist and his teacher's nemesis. He did not succeed in the latter endeavor, but did manage to embarrass Magendie for several years. The issue was recurrent sensitivity. Soon after Magendie reported to the Academy of Science the findings described in his 1839 lectures, Longet filed a claim for priority. Longet said that he had, in the course of one of Magendie's lectures and in the instructor's presence, performed an experiment that showed that the anterior root must receive its sensitivity from the posterior root. Magendie did not deny that Longet had made the experiment he described, but did deny that at that time Longet had drawn the conclusions he now claimed as his own. Only after Magendie had demonstrated the fact by different means did Longet come forward with his own story.[99]

The details of the case hardly matter, except as a revelation of Longet's animus and overweening ambition. Within a year he had reversed himself and was asserting as beyond doubt that the anterior roots are exclusively motor, the posterior exclusively sensory. If so, then Magendie's attribution of some sensitivity to the anterior root was an error, and the search for the source of that sensitivity futile. Longet's new tack proved more of a headache for Magendie than his priority claim. For if Longet's motives were questionable, his experiments were not. Operating on seventeen dogs, he had exposed ten nerve root pairs in each. Applying a variety of mechanical stimuli, he found the anterior roots to be insensitive, the posterior extremely sensitive. Galvanic stimulation of the anterior roots produced violent convulsion, whereas the same applied to the posterior roots produced none. In short, he had found no evidence of sensitivity in the anterior roots.[100] Further experiments, performed in the presence of professors Jean Cruveilhier and Pierre-Nicolas Gerdy of the Faculty of Medicine "and several French and foreign physicians," confirmed his results.[101] So did a commission named by the Academy of Science, though Magendie, who was a member, abstained from judgment. Longet was twice awarded the Academy's Montyon prize in experimental physiology for works that denied the existence of recurrent sensitivity.[102]

Recurrent sensitivity, in fact, went into temporary eclipse. Magendie and his assistant Claude Bernard were no longer able to demonstrate the property. Magendie later reported that his colleagues at the Academy treated the matter with a discreet silence, apparently following the maxim *Amicus Plato, sed magis amica veritas*.[103] Apart from the personal embarrassment to Magendie, the appearance of contradiction could only serve those who opposed a role for the experimental sciences in medicine.

Bernard, who had witnessed Magendie's 1839 experiments establishing recurrent sensitivity, returned to the problem at intervals in the 1840s. He began his own experiments in 1841. Though unsuccessful for several years, he was sustained by the idea that recurrent sensitivity was a transient property, the precise conditions of which had to be fixed. Only in 1844, when the anterior roots were generally regarded as insensible, did he recall the crucial difference between the 1839 experiments and subsequent ones. Magendie had prepared the experiment for his course in the morning. It was only later, after the animal had rested, that he had tested for the sensitivity of the anterior roots. In those circumstances the result was always positive. In the subsequent experiments of Magendie, Longet, and Bernard, in contrast, the anterior roots had been pinched immediately after exposure. Magendie and Bernard had even thought that promptitude was a condition of success. Bernard now conjectured that the reverse was the case, and that a period of rest following the operation to expose the roots allowed the animal to recover and the phenomenon to manifest itself. Bernard's first experiments along these lines were not successful, and preoccupation with other problems prevented him from pursuing the idea for the next two years. Only in 1846 and 1847 did he and Magendie finally succeed in determining the precise conditions under which recurrent sensitivity would or would not appear.[104]

Magendie published first.[105] Reflecting ruefully on the oblivion into which his discovery had fallen, he reaffirmed his 1839 findings and the physiological importance of the phenomenon. What he had overlooked, and what had contributed to the appearance of contradiction, was the possibility of a temporary loss of sensitivity in the anterior root. Such a loss does occur, he said, especially when the experimental animal is weakened by the pain and blood loss accompanying a laborious operation to open the cord and separate the root. In these cases it suffices to wait a few moments. The phenomenon will reappear, and will be maintained as long as the state of the wound and the surrounding parts permit.[106] Better still, he said, is to adopt a simpler, less traumatic procedure. The spinal cord is exposed on one side only, and to the extent of one or two vertebrae. Thus, "properly made" the experiment will yield an immediate positive test for recurrent sensitivity. The apparent contradictions in earlier investigations were a function of varying experimental conditions, and were therefore "only the rigorous expression of the facts."[107]

Bernard continued and expanded the argument. He had observed that recurrent sensitivity was especially pronounced in well-nourished animals in which digestion was proceeding. Fasting, he said, weakened and profoundly modified the property, while prolonged abstinence might make it disappear. Fasting and bleeding obviously acted in the same direction, by weakening the organism. In general, Bernard concluded, recurrent sensitivity required for its "full and entire manifestation" the presence of certain conditions: before the operation, vigor enhanced by nutrition; afterward, a sufficient rest to allow the denuded and cooled parts to reheat, and care not to pinch the posterior root before the anterior. Unfavorable conditions included a weakening of the organism by any cause, too great an exposure or cooling of the spinal cord, and any lesion of the posterior roots.[108]

The two collaborators published additional findings, and Bernard continued to refine the experimental procedures. The 1847 reports, to which were joined experimental demonstrations before scientific and medical colleagues, effectively silenced criticism of recurrent sensitivity.[109] More important for the future relations of science and medicine in France, they denied to medical critics of experimental physiology the use of what had been a prime example of the unreliability of experiment. Years later Bernard told his students at the Collège de France that recurrent sensitivity displayed the special difficulties of duplicating experimental conditions in physiology, in contrast to physics or chemistry. This was because of the "extreme complexity" of the organism's actions, and the "great mobility of its manifestations"—precisely those qualities that led some to speak of the inherent spontaneity and unpredictability of life processes.[110] Just because recurrent sensitivity was such a difficult case, its resolution was a significant vindication of the principle of experimental determinism. Bernard so used it in his most important statement on method.[111]

The research on recurrent sensitivity carried out by Magendie and Bernard between 1839 and 1847 was largely in the pattern established by Magendie in his 1820s work on the nervous system. Operative control of function by surgical intervention in the organism, accompanied by surgical language and instruments, played a prominent role.[112] Magendie, and later Bernard, underlined the importance of autopsy in experimental investigations.[113] The unspoken but obvious aim of the work was to illuminate human function, and comparative studies took little part even as means to that end. In these ways, as well as in the two physiologists' sensitivity to criticism of the reliability of experiment in their science, the work on recurrent sensitivity bears the imprint of the Paris medical milieu. Where it does not, and where it also departs from the pattern of Magendie's earlier research on the nervous system is in its infrequent reference to clinical or pathological experience. The displacement of this side of Magendie's program of pathological physiology is one expression of Claude Bernard's predominant role in these investigations. By the mid-1840s Bernard had moved away from medical practice. More important, he and other scientifically minded members of the Paris medical community were beginning to look beyond a medically oriented physiology to the ideal of a general science of life. The threads of continuity and change may be traced in Bernard's early career, and nowhere more clearly than in his research on the nervous system.

Anatomy and experiments on animals give a perfect account of morbid phenomena, and it is with the scalpel in hand, so to speak, that we explain the symptoms of paralysis of the face.

—*Claude Bernard* (1843)

9   From Medicine to Biology

Long before he passed his examinations for internship in the Paris hospitals in December 1839, Claude Bernard was on the path that would lead him to experimental physiology. He and Achilles Longet followed Magendie's course of lectures on the nervous system at the Collège de France in 1838–39.[1] By June 1839 he was acting as Magendie's assistant in lecture demonstrations.[2] After attending the services of Alfred Velpeau at the Charité, Jules Maisonneuve at the Hôtel-Dieu, and Jean Falret and Manec at the Salpétrière during his first year of internship, Bernard (with the support of the physician Pierre Rayer of the Charité) finally joined Magendie's service at the Hôtel-Dieu. By 1841, the second year of his internship, Magendie had made him his *préparateur* at the Collège de France.[3]

The highlights of the subsequent story are familiar enough. Taking his M.D. in 1843 with a thesis on gastric digestion, Bernard set up shop for himself with a private laboratory in the Commerce Saint-André-des-Arts. Unsuccessful as a private teacher of physiology and in attempts to obtain an academic post, his abandonment of science for medical practice was prevented only by a financially advantageous marriage to Marie Martin in 1845. Bernard remained in contact with Magendie. What began as a master-disciple relationship soon developed into collaborative research on a range of physiological problems involving the nervous system (including recurrent sensitivity), absorption, digestion, animal heat, and other topics. In 1847 Bernard was named *suppléant* to Magendie at the Collège de France, taking over the winter lectures from his teacher. On Magendie's retirement from the Collège in 1852, Bernard took on full teaching responsibility and was officially named to the chair on Magendie's death in 1855. By that time his major discoveries had been made, his reputation established. In 1854 he had been elected to the Academy of Science after winning its prize in experimental physiology on four separate occasions, and in the same year a chair of general physiology had been created for him at the Sorbonne.[4]

The first ten or twelve years of Bernard's scientific career were the most productive of important and original work in experimental physiology. The glycogenic function of the liver, the role of the pancreas in the digestion of fats, and the existence of the vasomotor nerves are all discoveries whose essentials date from the period 1841–1853. The beginnings of Bernard's studies of the action of poisons and their use in physiological analysis also date from those years.[5]

Bernard has been the object of an impressive scholarship that has revealed much about his milieu, his scientific personality, and his achievements.[6] Detailed reconstitutions of Bernard's work on carbon monoxide and curare poisoning by Mirko Grmek, and on digestion and nutrition by Frederic Holmes, have documented the extent of Bernard's debts to predecessors and contemporaries, especially Magendie. The same studies have exposed the gaps between Bernard's published accounts of his work, which are often rational reconstructions that condense, select, or simplify the process of discovery, and that process itself, replete as it usually was with groping and false starts. Grmek and Holmes have also shown how the general physiological and methodological ideas for which Bernard is famous emerged gradually from Bernard's reflections on his own experimental practice. Both have emphasized that it was only in the second decade of his scientific career, in the 1850s, that Bernard began to display the philosophical self-awareness and broadly biological orientation that were to mark his later work.

Less well recognized is the extent of Bernard's indebtedness to the Paris medicine of the first half of the century. Bernard's best-known work, *Introduction à l'étude de la médecine expérimentale* (1865), has done much to shape our image of the Paris clinical school and its relations to the experimental sciences. Bernard saw a positive role for clinical and pathological experience when critically weighed in conjunction with experimental investigation. On the whole, however, he represented his science in contrast with and struggling against the prevailing tendencies of contemporary medicine. In Bernard's eyes his physiology was active, interventionist, and based on the notion of experimental determinism. Paris clinical medicine, in contrast, was passive, observational, and empiricist, tending to acquiesce in the notion of the inherent spontaneity of the phenomena of life and the limitations it placed on the scientific investigation of the organism. This image of contrast and opposition has been taken up and developed by leading scholars of the period. Ackerknecht has underlined the resistance of clinicians to the experimental sciences, the outsider status of experimentalists like Magendie in the Paris medical community, and the "dead end" in which hospital medicine found itself by the 1840s.[7] Georges Canguilhem emphasizes Bernard's opposition to the Hippocratic, natural history (nosological), and pathological anatomical approaches to medicine on the grounds that they were passive, contemplative, and descriptive. He highlights Bernard's view that experimental medicine would ultimately make nosology and pathological anatomy superfluous, because it would show diseases to be not distinct entities but distorted physiological functions. With pathology a branch of experimental physiology, the way would then be open for the rational control of disease through a new therapeutics.[8] The "truly promethean idea" of experimental medicine and physiology that pervades the *Introduction* is, in Canguilhem's view, its most original feature. Through its ethos of power over nature, Bernard's experimental medicine effected a clear break with previous medical thought and practice. In so doing it became "one of the figures of the demiurgical dream dreamed by all the industrial societies of the mid-nineteenth century, in an age when, through their applications, the sciences had become a social power."[9]

Such contrasts, though they accurately mirror Bernard's own later thought, are less than adequate as characterizations of his relations to the milieu in which his science took form. Written in the third decade of his career, when Bernard was still struggling with some elements of the scientific and medical communities over the definition and permanent status of his field, the *Introduction* is a polemical work, though an exceptionally thoughtful one. It is directed against the limitations and prejudices of Paris clinical medicine vis-à-vis experimental science. It embodies oppositions that were much clearer to Bernard in the 1860s than they were in the initial and most creative period of his scientific work. Looking back on his early discoveries, Bernard simplified the relation between the "chemical" and the "physiological" approaches in the events that led to his discovery of the source of animal sugar, in so doing producing a didactic rationalization of those events.[10] In a similar fashion he simplified the relations between experimental science and clinical medicine in the interests of a more dramatic defense of the former. That Bernard felt the need to write such a work speaks for the persistent reality of the limitations and prejudices he attacked. It should not obscure the complex, subtle, and sometimes profound ways in which physiology was shaped by its medical environment over the half century that culminated in Bernard's early career.

Suggestions and fragments of evidence for the medical background of Bernard's physiology are available in the literature. In one of his last publications, Joseph Schiller noted the potential interpretive significance of the practical training of the physiologist, his unspoken manual or craft technique.[11] Holmes emphasized the role played by Bernard's "extraordinary surgical skills" and "alertness to operative opportunities" in his discovery of the source of sugar in the animal economy.[12] Grmek has called attention to Bernard's consistent use of physiological autopsy in his experiments on curare, his reference to poisons as "chemical scalpels" or "means of elementary vivisection," and his concept of poisoning as the production of a local lesion in his early investigations on poisons.[13] All accounts have represented Bernard's ultimate aim of basing pathology, and through it therapeutics, on experimental physiology. These points and others, when pursued and consolidated, define a medical dimension of Bernard's physiology that underlies all his particular investigations. Shaped by his professional training, his apprenticeship with Magendie, and his personal connections with the Paris medical community, it finds expression in the surgical qualities of Bernard's experimentalism, his use of autopsy and of clinical and pathological material, his localism, his concentration on human physiology, and his aim to create a scientific medicine.

In the first decade of his scientific career, Bernard's physiology underwent an important shift in its relation to his medical origins. Bernard moved from a coordinated use of experiment and clinical or pathological experience to a clearer emphasis on the subordination of pathology to physiology. This movement was reflected in and reinforced by the program of the newly formed Société de Biologie of Paris, in which Bernard served as one of the first vice-presidents. Even with this change of emphasis and with the subsequent biological and philosophical broadening of Bernard's interests, however, his physiology retained the marks of its ancestry in the Paris clinical and scientific world of the early nineteenth century. So too did the institution that encouraged in Bernard and others of his generation a more expansive conception of his science. In its stated aims, its membership, and its early publications, the Société de Biologie embodied its medical origins even as it reached beyond them to the ideal of a general science of life.

Nowhere are the continuity between Magendie and Bernard and the beginnings of the transition from the medical to the biological ideal in physiology more clearly displayed than in Bernard's investigations on the nervous system between 1843 and 1853. The 1840s were years of quickened interest in nervous phenomena. Pierre Flourens, whose research in this area had begun almost simultaneously with Magendie's, was still very active. The second edition of his monograph on the nervous system, sometimes critical of Magendie, appeared in 1842.[14] Achilles Longet, a graduated M.D. teaching a course in experimental physiology at the Ecole Pratique, had already singled out Magendie's work on recurrent sensitivity for attack in 1839. His *Anatomie et physiologie du système nerveux,* published in 1842, also took to task the views of his former teacher at the Collège de France.[15] In a long memoir one year earlier, Longet had singled out the spinal cord and spinal roots as his special objects of interest.[16] The mid-1840s also witnessed the beginnings of the scientific career of Charles-Edouard Brown (later Brown-Séquard). Brown's M.D. thesis, published in Paris in 1846, reported the results of numerous animal experiments on the spinal cord and included frequent references to the work of Marshall Hall, Magendie, and Longet.[17]

Still more symptomatic of interest in the nervous system in the Paris medical and scientific community was the founding of a new journal, *Annales médico-psychologiques,* in 1843. Its editors were all medical doctors: Jules Baillarger, physician to the insane at the Salpêtrière; Laurent Cerise; and Magendie's critic and rival, Achilles Longet. Their primary intention, stated in the introduction to the first volume, was practical and medical: to advance the understanding of nervous phenomena in their relation to disease. The editors recognized four categories of nervous pathology: mental illness per se (*aliénation mentale*), which did not necessarily involve physical lesion of the nervous system; the legal medicine of mental illness; nervous disorders (*névroses*) such as hypochondria, hysteria, catalepsy, epilepsy, somnambulism, and neuralgias, which presented "a complex of phenomena to which all parts of the nervous system seem to contribute"; and organic alterations of the nervous system.[18]

Mental illness per se had been the principal object of Pinel's program, and the editors felt that Pinel and Jean Esquirol had established such firm foundations for its study that they could afford to widen the scope of their endeavor to include the other categories. Legal medicine, of course, was an administrative or professional rather than a scientific division. The nervous disorders were to be illuminated by a new science of the relation of the physical and the moral, which would investigate the interaction of physiological and psychological phenomena. In this the editors clearly looked for inspiration to Cabanis, whose name they cited and whose *Rapports du physique et du morale de l'homme* was just being republished with an interpretive essay by Cerise. Referring to the ongoing philosophical debate between spiritualists and materialists, they pledged to accept worthy contributions from partisans of both schools.[19]

Regarding the last category, organic alterations, the editors were oddly equivocal. On the one hand, they recognized and honored the impulse given to study of the structure and function of the nervous system "about a half century ago," and the persistence of that impulse in the 1840s. They fully appreciated the illumination that the organic alterations of the nervous system had received from "the principle of localizations" as it had been applied in experimental physiology and pathological anatomy. Diagnosis had thereby achieved "a truly remarkable precision," and there had thus occurred "an important exchange of services between the clinic and physiology." Yet they held that not only could the principle of localizations not be applied to the study of nervous disorders (*névroses*), but the impulse that sustained the study of the structure and function of the nervous system "was too great not to have a tendency to exhaust itself in the near future." Therefore, they could not promise their readers many works of that type.[20]

On the last point the editors could hardly have been more wrong. With unintended irony, and as if to underline their mistake, they included in the same initial volume of the *Annales* the first published scientific paper of Claude Bernard, one that examined the role of alteration of the chorda tympani nerve in the production of facial hemiplegia. Viewed in conjunction with his paper of the following year on the alteration of the sense of taste in facial paralysis, this article was revealing of the extent to which Bernard's early physiology was in the mold of his teacher's.[21]

In introductory remarks Bernard noted that although the chorda tympani had been known to anatomy since the sixteenth century, its precise functions were still obscure and subject to contradictory hypotheses. The vicious circle of hypothesis and counterhypothesis would be broken only by recourse to anatomy and experience, in the form of animal experiment, comparative anatomy, and pathology. Comparative studies had the special merit of correlating differences in function with differences in organization. Nature thus performed "veritable experiments" that were all the more valuable for avoiding the mutilations that often complicated vivisections. Pathology would be illuminated by the results of physiology.

Turning first to anatomy, Bernard described with great care and exactness the course and connections of the chorda tympani nerve in humans (see Figure 1): its origins in the facial nerve (seventh cranial pair), its association with the lingual nerve (mandibular branch of the trigeminal or fifth cranial nerve) before this nerve gave off any terminal branches, and its termination in the tongue. He also described the submaxillary ganglion and its connections with both the lingual nerve and the submaxillary and sublingual glands. He used a variety of dissection techniques including mac-

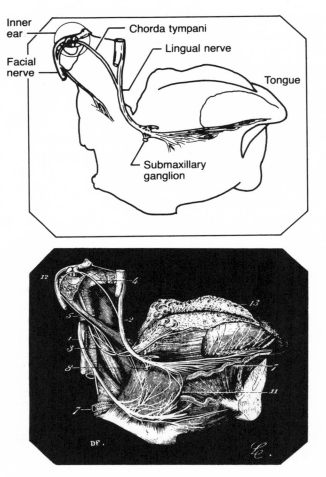

(1) *Organe du goût.* (Figure empruntée à la *Nevrologie* de MM. Lu-
dovic Hirchfeld et Leveillé). — **1**, grand hypoglosse; — **2**, branche
linguale du trijumeau ; — **3**, branche linguale du glosso-pharyngien; —
4, corde du tympan ;—**5**, rameau lingual du facial qui, après s'être anas-
tamosé avec le glosso-pharyngien, parvient à la langue ;— **7**, plan nerveux accompagnant l'artère linguale et sa division ; — **8**, ganglion sous-
maxillaire donnant des ramifications à la glande sous-maxillaire ; —
11, anastomose du nerf lingual avec le grand hypoglosse ;— **12**, nerf facial; — **13**, épiderme détaché du derme et déjeté en haut.

Figure 1. The chorda tympani nerve in humans. Above: schematic drawing. Below: original from Claude Bernard, *Leçons sur la physiologie et la
pathologie du système nerveux* (Paris, 1858).

eration, solvents, and microscopy. Cadavers were probably available to Bernard through Magendie's service at the Hôtel-Dieu, though he did not specify the source of his dissection material.

On the basis of these anatomical considerations alone, Bernard concluded that the chorda tympani constituted the motor element of the lingual nerve. Implicit in this conclusion, however, was Bernard's commitment to Magendie's (mistaken) view that the facial nerve was purely motor in function. Comparative studies showed that the chorda tympani nerve existed only in mammals. With its disappearance in birds and reptiles the lingual nerve properly speaking, and thus the faculty of taste, also disappeared.[22]

With anatomical connections clearly displayed, experiments could be undertaken to determine the functions of each component of the lingual nerve. Bernard took as given Magendie's finding that the chorda tympani was insensible to pinching or sectioning, and also Magendie's general view that the nerves of the special senses were insensible to mechanical irritations. On this point he found that the lingual nerve resembled the nerves of the special senses.

Because of the complex trajectory of the chorda tympani in the head, experiment was difficult. As the physiologist Pierre Bérard had noted, the nerve seemed to be placed as a deliberate challenge to the sagacity of physiologists. Bernard pointed out that something could yet be learned by sectioning the facial nerve at its origin, before its anastomosis with the vidian and pneumogastric nerves.

In the first experiment, Bernard anesthetized and immobilized an adult dog with a strong dose of opium extract. He introduced a small hook with a double cutting edge (*à double tranchant*) into the skull on the left side via the orifice of the mastoid vein situated above and within the mastoid apophysis. As soon as the instrument had penetrated, Bernard directed it obliquely down and inside, following the posterior face of the petrosal bone. As soon as contractions were visible on the left side of the face, he knew the instrument had reached the facial nerve. Turning the hook upward, and without leaving the petrosal bone, he carefully withdrew the instrument, thereby pulling at and sectioning the nerve. The completion of this operation was signaled by the immediate and complete paralysis of the left side of the face. Within six days the wound had healed and the effects of the opium had dissipated. Bernard was able to confirm that, apart from the facial paralysis, there was a considerable diminution of the gustatory faculty in the left anterior half of the tongue, without any corresponding alteration of movement or of the tactile sense in the same region. When the animal was sacrificed after thirty-three days, autopsy confirmed that the seventh pair, and only the seventh pair, of cranial nerves had been sectioned. He obtained the same results in experiments on two other dogs.[23]

Bernard concluded that the chorda tympani originated in the primitive fibers of the facial nerve and that, as a motor nerve, the chorda tympani through its anastomosis with the lingual nerve exerted a direct influence on the sense of taste in those regions of the tongue to which the lingual nerve was distributed.[24]

Bernard knew that in the opinion of some researchers the vidian nerve contributed sensitive fibers to the chorda tympani. If so, sectioning the latter nerve should yield results different from those obtained in the first experiment. To rule out this possibility, Bernard devised another experiment in which the chorda tympani was sectioned after its emergence from the facial nerve—that is, after the facial had already been joined by the vidian and pneumogastric branches. The operation was performed on the middle ear of an adult dog, where the nerve is found free and isolated and can most easily be pulled or cut. He introduced a small hook, piercing on its convex side and with a cutting edge in its concavity, into the external auditory canal. A slight resistance and a peculiar cracking sound (*un craquement particulier*) signaled the penetration of the tympanic membrane. Bernard immediately turned the hook upward and drew it out, pulling and sectioning the nerve. Taking his own superior operative ability for granted, Bernard described the procedure as simple and infallible, and could not understand why Luigi Guarini had abandoned it as too difficult. Repeated on eight dogs, and confirmed in each case by autopsy, the procedure yielded physiological results identical to those of the first experiment. Bernard concluded that the vidian contributed nothing to the chorda tympani, and that the latter nerve had no action on the tactile sensibility of the tongue.[25]

The anatomist Friedrich Arnold had found fibers from the chorda tympani supplying the submaxillary ganglion. Since there were also fibers emanating from this ganglion and associated with the walls of Wharton's duct, he had concluded that the contractions of this duct were under the influence of the chorda tympani. Against this conclusion Bernard cited the results of comparative, experimental, and pathological studies. The Wharton's ducts of the horse were very large, yet these animals possessed no submaxillary gland, properly speaking. Sectioning the chorda tympani "in a large number of dogs" yielded no observable alteration of salivary secretion. Clinical cases of complete paralysis of the facial nerve were never associated with dilation of Wharton's ducts. Anticipating yet another objection—that the chorda tympani mediated an interaction between the sense of taste and the secretion of saliva—Bernard diversified his experimental response. On two dogs he removed the submaxillary glands altogether; on others he tied the Wharton's ducts below the jaw; on still others he sectioned these ducts and drew them out, creating an exterior salivary fistula. In no case was there appreciable modification of the sense of taste, a finding consistent with clinical observation of diseases or extirpations of the submaxillary glands.[26]

Bernard felt that the chorda tympani acted as a motor nerve by somehow speeding up the perception of taste by the tongue, but he was anxious to exclude any role for the nerve in the tongue's movement. This position involved a critique of the work of Guarini, which did assert such a role. Three points of this critique are of special interest. First, Guarini's operative procedure had involved cutting the chorda tympani in the temporal-spinal fossa (*la fosse temporo-épineuse*) at the point where this nerve joined the lingual nerve. This required a large wound, detachment of the pterygoid plates, division of many large blood vessels, and the exposure of several important nerves including the branches of the inferior maxillary nerve. Not surprisingly, the first two animals had died of hemorrhage, while the third had been described by Guarini as more dead than alive after the operation. In Bernard's view, results obtained through such extensive mutilations, and on a single animal, were not to be trusted. Second, the independence of peristaltic movements of the tongue from control of the chorda tympani was demonstrated by experiments in which Bernard sectioned this nerve on one side in two dogs. After several weeks, the dogs were sacrificed with hydrocyanic acid. The fibrillary movements of the tongue ordinarily observed at the moment of death by this poison were seen on both sides—that is, regardless of whether or not the muscle tissue was supplied by the chorda tympani. Finally, criticizing Guarini's view of the role of the chorda

tympani in voice production, Bernard cited comparative observations not only on dogs and cats but also on birds—which, he
pointed out, were capable of a wide range of vocal modulation
though lacking a chorda tympani nerve.[27]

Pathological observation could now be given its due. "Anatomy
and experiments on animals," Bernard confidently asserted, "give
a perfect account of morbid phenomena, and it is with the scalpel
in hand, so to speak, that we explain the symptoms of paralysis
of the face." So far, such methods had given an adequate account
only of the more superficial nerves, however, "and it is not surprising that pathology, following the progress of physiology, should
remain impotent as long as this assistance was lacking." Bernard
described several clinical cases in which facial hemiplegia was accompanied by loss of the sense of taste in the anterior portion of
the tongue on the affected side. The latter symptom was not invariably present, however, and he speculated that its presence or
absence depended on whether the nerves were affected in the
periphery or at a deeper level. For the moment, he would follow
the general rule that physiological results and pathological symptoms "lend themselves a mutual support, that one may be interpreted by means of the other." Thus, he could be sure that the
alterations of taste observed in facial hemiplegia depended on
paralysis of the chorda tympani.[28]

In several ways, this first published paper by Bernard replicated
patterns already established by his teacher. Each section began
with the recounting of previous opinions and the promise to resolve their uncertainties by direct experiment, a formula for Magendie's presentations. Bernard explicitly advocated a coordinated
approach that included animal experiment, comparative anatomy
and physiology, and clinical and pathological experience. More
than that, he made use of all three methods, and related them to
one another. Yet, as for Magendie, comparative studies remained
means to the end of the extension of human physiology and human
pathology. In its use of anesthesia and surgical instruments, its
dependence on exact anatomy and manual dexterity, its language
of "operations," "procedures," and "autopsies," and its aim to control the manifestations of physiological and pathological phenomena, Bernard's experimental work, like Magendie's, displayed a
strikingly surgical character.

Yet even in this first independent effort subtle divergences may be discerned. Bernard's greater versatility in the design of different experiments around the same problem was already apparent. His use of the known effects of a specific poison on a specific muscle to test the action of a specific nerve revealed an awareness of the potential usefulness of drugs in physiological analysis that he would later implement with great effectiveness. And although Bernard followed his teacher in recognizing the mutual illumination that physiology and pathology may provide one another, there was already a tangible emphasis on the leading role of physiology— an emphasis that was to become the dominant one in his later thinking.

This was not the end of the story. The following year Bernard published a sequel to the first article, filling out the pathological and clinical side of his problem and venturing a step further in its theoretical explanation. Pertinent clinical observations were supplied by Pierre Rayer and Velpeau at the Charité, François Guéneau de Mussy at the Enfants Malades, and Auguste Chomel at the Hôtel-Dieu.[29] One example, taken from Rayer's service at the Charité, must suffice. A thirty-five-year-old man entered the hospital on 29 June 1843 with clear symptoms of advanced tuberculosis of the lungs. Before coming to the hospital the patient had had a flow of pus from the left ear, and one day a small bone was expelled with it. Paralysis of movement in the entire left side of the face followed. Sensibility remained intact, except that the sense of taste was greatly weakened on the left side of the tongue. Rayer attributed the facial paralysis to a lesion of the seventh pair of nerves consequent on a tuberculous affection of the petrosal bone. The patient died of pulmonary phthisis on 15 December 1843. At Rayer's request, Bernard attended the autopsy and carefully examined the pathological alterations relative to the left facial nerve. He found that the facial nerve was affected by tuberculous degeneration in the canal of the petrosal bone, and disappeared in the softened tuberculous mass of the middle ear. He discovered no lesion in the branches of the trigeminal and lingual nerves. The salivary glands and their conduits were also intact on both the right and left sides.[30]

On the basis of this and similar cases, Bernard concluded that loss of the sense of taste may or may not accompany facial paralysis, depending on whether or not there is a lesion of the chorda tympani. He pointed out that the symptom of loss of taste had not been noticed by pathologists, or had been regarded by them as exceptional, probably because it had been overshadowed by the disorders of the facial muscles. Once again, therefore, the emphasis fell on physiology as the guide for pathology.

Physiology, however, still could not say by what mechanism the chorda tympani affected the sense of taste. Bernard's explanation, based on his view of the nerve as purely motor, was that it acted by influencing the papillae of the tongue. Placed between the epithelium and the terminal branches of the sensory nerves, the papillae, Bernard thought, necessarily mediated the contact between sensory excitants and those nerves. In this their action could be compared with the intestinal villae. In a spontaneous generalization characteristic of his research thinking, Bernard concluded that "the papillary network of the tongue is found to be completely analogous to the modifying apparatuses placed between the other nerves of sense and their natural excitants."[31]

Bernard had therefore followed Magendie's example in yet another respect, by seeking out clinical and pathological observations that would simultaneously confirm the experimental results and be explained by them. Like Magendie, he could depend on the sympathetic cooperation of a number of Paris physicians, including several leading hospital clinicians. The routine of pathological anatomy, the correlation of clinical symptom with anatomical lesion through autopsy, came easily to Bernard, a recent product of the Paris medical faculty and a veteran of an internship in the Paris hospitals.

By the fall of 1842, simultaneous with or even prior to the beginning of his work on the chorda tympani, Bernard had begun studies of the spinal or accessory (eleventh cranial) nerve. Published in the *Archives générales de médecine* in 1844 and awarded the Academy of Science prize in experimental physiology the following year, this research is even more revealing of the continuity of Bernard's early work with that of Magendie, both as an expression of antisystematic, empiricist attitudes and as a clear case of the essentially surgical character of physiological practice.[32]

Reviewing earlier opinions on the functions of the spinal nerve, Bernard singled out those of the German physiologist Theodor Bischoff for special commentary. Writing in 1832, Bischoff had asserted that the two cranial nerves, the vagus (or pneumogastric) and the spinal, were to each other as the two roots of a spinal pair, the vagus being the sensitive root and the spinal the motor root. Doubts had been raised by Müller, Magendie, and Bérard, but Longet and others had been convinced that this view was correct. The problem, as Bernard saw it, was that Bischoff had leaned too heavily on the analogical appeal of an extension of the Bell-Magendie law to the cranial nerves. Rather than varying his procedures in the light of different alternatives, he had simply sought confirmation for a prior conviction. He had therefore been prepared to interpret his single reasonably reliable experiment as demonstrative. In words that might have been Magendie's, Bernard remarked that "it would not be the first time that even in starting with a true fact, reasoning in physiology led us to a system instead of to the truth."[33]

Anatomy alone belied Bischoff's conclusion. The spinal nerve originated from two sets of tributary filaments, one arising from the medulla oblongata, the other from the cervical region of the spinal cord (see Figure 2). These joined together in a single trunk just below the vagus (pneumogastric). This trunk soon divided into two branches: a smaller internal one, which on maceration was revealed as the continuation of the fibers arising from the medulla oblongata; and a larger external one, which the grossest dissection showed to be continuous with the fibers coming from the cervical region. The internal branch joined the vagus and subdivided into two quite distinct branches. One of the latter contributed to making up the superior pharyngeal branch of the vagus, whereas the other disappeared into the trunk of the vagus not far from the origin of the latter's superior laryngeal branch. The external branch of the spinal nerve remained largely distinct from the vagus, receiving only a few filaments sent to it from that nerve. It was then bent down and outside and, after associating with nervous filaments from the cervical plexus, terminated in the sterno-mastoidian and trapezius muscles. The key point in the context of Bernard's critique of Bischoff was that in spite of their anastomoses, the spinal and vagus nerves remained anatomically distinct. They did not stand in relation to each other as the two roots of the same spinal pair, which invariably came together to form a single nerve and were then inseparable even by maceration.[34]

The deficiencies of Bischoff's anatomical analysis were not made good by his experiments. Operating on five dogs and two young

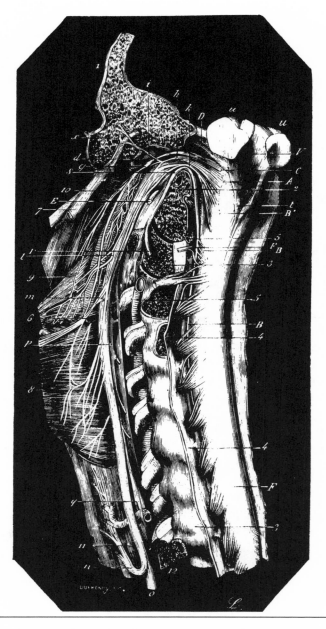

Figure 2. Origins of the spinal or accessory (eleventh cranial) nerve in humans. Upper right: schematic drawing. Above and lower right: original from Claude Bernard, *Leçons sur la physiologie et la pathologie du système nerveux* (Paris, 1858).

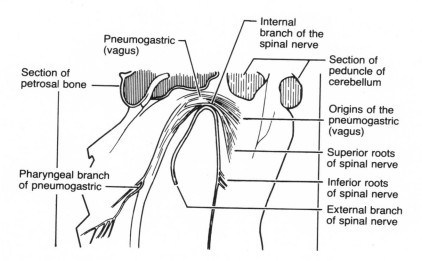

Internal
branch of the
spinal nerve

Pneumogastric
(vagus)

Section of
peduncle of
cerebellum

Section of
petrosal bone

Origins of the
pneumogastric
(vagus)

Superior roots
of spinal nerve

Inferior roots
of spinal nerve

Pharyngeal branch
of pneumogastric

External branch
of spinal nerve

(1) *Nerfs pneumogastrique et spinal chez l'homme.* — La pièce, vue en arrière, a été disséquée et disposée de manière à mettre en évidenc les origines et les anastomoses de ces nerfs. — A, faisceau des origines du pneumogastrique ; — B, filets originaires de la grande portion médullaire du spinal qui vient ensuite former la branche externe de ce nerf *r* ; ces filets originaires s'étendent depuis la première jusqu'à la cinquième paire cervicale environ ; — B', filets originaires de la portion bulbaire du spinal qui vont ensuite constituer la branche interne de ce nerf *k* ; — C, origine du glosso-pharyngien ; — D, troncs du facial et de l'acoustique réunis après leur origine (septième paire) ; — E, nerf grand hypoglosse coupé ; — F, F, racines postérieures des paires nerveuses cervicales rachidiennes ; — *g*, ganglion du nerf glosso-pharyngien ; — *h*, ganglion jugulaire du pneumogastrique ; — *i*, rameau auriculaire du pneumogastrique ; — *k*, branche interne du spinal ; — *l*, rameau pharyngien du pneumogastrique provenant de la branche interne du spinal ; — *m*, nerf laryngé supérieur ; — *n*, nerf laryngé inférieur ou récurrent ; — *o*, tronc du nerf pneumogastrique coupé ; — *p*, ganglion cervical supérieur ; — *q*, ganglion cervical inférieur ; — *r*, branche externe du nerf spinal coupé ; — *s*, anastomose de Willis entre le pneumogastrique et la branche externe du spinal ; — *t, calamus scriptorius* ; — *u, u*, coupe des pédoncules du cervelet ; — *v*, plancher du quatrième ventricule ; — *x*, corde du tympan ; — 1, coupe du rocher ; — 2, coupe de la partie basilaire de l'occipital ; — 3, 3, vertèbres cervicales ; — 4, 4, dure-mère ; — 5, 5, artère vertébrale ; — 6, 6, artère carotide ; — 7, faisceau des muscles styliens coupés ; — 8, 9 et 10, muscles constricteurs du pharynx ; — 11, œsophage ; — 12, première vertèbre dorsale.

goats, Bischoff had attempted to section all the roots of the spinal nerve, expecting that all motor activities controlled by the spinal nerves, such as the voice, would be abolished. The procedure consisted of two steps. Bischoff exposed and divided the fibrous membrane uniting posteriorly the occipital (base of the skull) and the atlas (or highest vertebrae, joined to base of skull). This laid bare the vertebral cavity with the two spinal nerves on either side of the medulla oblongata. It also enabled the operator to section the inferior tributaries of the spinal nerve. In dogs and goats, however, it did not permit section of the superior tributaries. To accomplish this, a second step—the removal of a portion of the occipital—was required. Whereas the first step was relatively easy—in fact identical to Magendie's technique for obtaining cerebrospinal fluid—the second almost inevitably involved sectioning the venous sinus and causing profuse bleeding. In only one of seven trials had this method yielded unambiguous results, and Bischoff had not repeated this success. Similarly, Longet had been unable to repeat Bischoff's experiment, succeeding only in altering the voice of his experimental animals. Nevertheless, Longet had drawn even more sweeping conclusions, asserting that the internal anastomosis of the spinal nerve controlled all movements of the larynx, trachea, bronchae, pharynx, esophagus, and stomach.[35]

To Bernard, the single experiment that supported this imposing edifice was rendered of still more dubious value by the extensive mutilations of Bischoff's procedure. Not only were the medulla oblongata and the cerebellum exposed, but the animals' lives could not be maintained. They perished either in the course of the operation or shortly thereafter. Ostensibly the cause was loss of blood, but in repeating the experiment many times Bernard had noted that in most cases death was due to the introduction of air into the heart via the venous sinus or the small veins of the skull. His own attempts to avoid this by tying the four jugular veins or by obstructing the superior vena cava proved unsuccessful: death was slowed, but autopsy showed that it occurred in the same way.

Cauterization, which directly closed the occipital sinus, was slightly more successful. In one procedure, a knife made red hot by fire served simultaneously to section the occipital and cauterize the wound. In a different operation, Bernard used a perforator to make a small hole just below the external occipital line. He introduced the conical siphon of a small syringe by means of this opening, and injected a concentrated solution of persulfate of iron or of silver nitrate into the sinus. Done conjointly with the temporary ligature of the jugulars, this method usually obstructed the sinus by coagulating its blood. By these ingenious but difficult modifications of procedure, Bernard was able to secure a few extra hours of life for three dogs and one cat, while exposing the inferior and superior roots of the spinal nerve in all cases. Sectioning these roots progressively, he found that only the division of the superior roots affected the voice. When all the superior roots of the spinal nerve were cut, the voice was completely lost. This result was consistent with that indicated by his anatomical analysis. Still, the mutilations were such that the animals remained very ill, and none survived more than five hours. He could draw no reliable conclusions regarding the effect of the spinal nerve on the lungs or other organs supposed to be under its control.

If that was to be done, a wholly new procedure was required. This Bernard found, and its centrality to his entire investigation of the spinal nerve justifies a detailed description. The experimental subjects were dogs, cats, rabbits, and rats. An extended incision from the mastoid apophysis to a little below the transverse apophysis of the atlas exposed the external branch of the spinal nerve at the point at which it disengaged behind the sternomastoidian muscle. Bernard directed an aide to raise the superior part of this muscle with a small tenaculum. Taking precautions to avoid the neighboring nerves and vessels, Bernard continued the dissection, following the nerve to its entrance to the skull at the jugular foramen (*trou déchiré postérieur*). He then grasped the isolated nerve with pincers specially modified for the operation. In a footnote, Bernard explained that this instrument resembled closely the torsion pincers for arteries, except that the serrations that comprised the bit (*mors*) had to be rounded rather than incisive so that the nerve could be firmly grasped without cutting into its substance. He exerted a continuous traction. A cracking sound (*craquement*) with simultaneous loss of resistance was followed by the withdrawal of a long conical nervous cord, ending in a very tenuous extremity. This, Bernard concluded, was nothing less than the entire intracranial portion of the spinal nerve. Such a finding could be confirmed only by autopsy, and he carried out this procedure in all cases. In fact, Bernard reported that before trying the operation on living animals he had carefully studied its effects on dead ones in which the origins of the spinal nerve had been exposed. In living subjects the wounds of the neck healed after four or five days and the animals were in their normal state except for loss of the spinal nerve.[36]

The method was simple, reliable, and avoided mutilations. As an operative procedure it was easy in all cases except the dog, where the connective tissue joining nerve to bone was generally too dense to allow extirpation. The operation had never produced complications. It could be—and was—applied to other nerves: the facial, the hypoglossal, and the cranial nerves in general. Best of all in the case of the spinal, it permitted the investigator to isolate and study at leisure the effects of ablation of that nerve.

Bernard's results were not favorable to Bischoff's theory. In the active animal, sectioning the spinal nerves produced loss of voice, inefficient swallowing, brevity of expiration when the animal attempted to make a loud sound, panting in large movements, and sometimes irregularity of locomotion. In the resting animal, all of the organic, respiratory, circulatory, and digestive functions occurred as though the spinal nerves were still intact.

The most striking and constant symptoms—voice loss and trouble in swallowing—were clearly due to paralysis of movements of the larynx and pharynx. Yet the effect on these organs of sectioning the spinal nerves was quite distinct from that produced by loss of the laryngeal nerves, which were branches of the vagus. Whereas the former operation resulted in permanent dilation of the glottis, the latter produced occlusion of the same organ. The effects of the spinal and laryngeal nerves on the larynx were thus distinct and antagonistic. The first controlled movements of phonation and deglutition; the second, movements of respiration. It was a case in point of the lack of a one-to-one relation between structure and function, since, as Bernard argued, the two nerves did not influence distinct sets of muscles in the larynx but brought distinct influences to the same muscles. Moreover, such a double innervation of a single muscle was not an isolated case "but a general means that nature employs to economize on the number of motor organs."[37] The attribution of a motor function to the vagus that did not derive from the spinal nerve was alone sufficient to invalidate Bischoff's view of the vagus as the sensitive root of a spinal pair.

The other symptoms—brevity of expiration, panting during movement, irregularity of locomotion—resulted from sectioning the external branch of the spinal. This nerve supplied the sternomastoid and trapezius muscles, which in turn played roles in respiration and locomotion. Again the situation was not a simple one, however, since in normal inspiration the same muscles were controlled by the cervical plexus.[38]

Taking the work on the spinal nerve as a whole, a clearer exemplification of the surgical character of Bernard's physiology could scarcely be imagined. Much of the investigation hinged on the invention of a new experimental or—as Bernard more often called it—operative procedure. The terminology in which the application of that invention was described—"procedure," "operation," "autopsy"—was itself borrowed from surgery and from the clinic. Not only the language but the substance as well reflected that origin, though the immediate aims in this case were scientific rather than therapeutic. Again and again Bernard displayed and deployed an intimate and meticulous knowledge of gross anatomy and a sure confidence in dissection. His instruments—scalpel, cautery, syringe, perforator, tenaculum, and pincers—were borrowed from the operating room, though sometimes, as in the case of pincers used to extract the spinal nerve, they were modified for specific ends. Where necessary, he called on an aide for assistance in their use. Even through the verbal account, we sense a consummately skillful craftsman at work and can only wonder at the loss to surgery entailed by Bernard's scientific vocation.[39]

All the more remarkable, therefore, was the relative absence of pathological experience from this investigation. Oblique references were there, and no doubt Bernard was sensitive to the implications of whatever material came his way.[40] But its role was peripheral.

Scarcely less so was his use of comparative anatomy and physiology. Again, the references were there, and on several occasions their use was quite substantive. In support of his position on the distinctness of the vocal and respiratory functions of the larynx, for example, Bernard pointed out that in birds the vocal larynx was completely separated from the respiratory larynx. In arguing that the external branch of the spinal nerve controlled the sternomastoid and trapezius muscles, which in turn controlled the movements of the thorax in respiration, Bernard noted that in birds the thorax was immobile, due to its bony structure. Thus no external branch of the spinal nerve was needed, and none existed.[41] Yet though they were not exactly peripheral, the most that can be said for these references is that they played an ancillary and supportive role in the overall argument. Center stage was reserved for experiments performed on a limited and for the most part familiar group of mammals: dogs, cats, rabbits, goats, and so on.

Bernard's posture in these memoirs, if not empiricist, was at least strongly antisystematic. In this it was a faithful echo of Magendie. It was also—as far as it went—consistent with Bernard's later position as it evolved in the 1850s. Yet one element of that later position was still conspicuously absent. Bernard evidently did not yet see a positive role for hypotheses or preconceived ideas in research. If he had, he would have attributed a qualified positive value to Bischoff's theory, though that theory was later proved fallacious. This he did not do. At the same time, apart from his formal methodological position, another and in some ways divergent trend in his thinking was evident. This was his irrepressible tendency to physiological generalization. Here the occasion was provided by the dual and antagonistic innervation of the muscles of the larynx by branches of the spinal and vagus nerves. This, Bernard stated with assurance, was but one instance of a general mode of operation of nature. Indeed, other instances would later be found—the multiple functions of the liver would be a more famous example—but in the present context it was an assertion that belied the stern empiricism of its author.[42]

For the most part, Bernard's investigations on the nervous system in the eight years after 1844, up to and including his early work on the vasomotor nerves, continued the patterns established in these early memoirs. Animal experiment, mainly study of the effects of ablation of nerves, remained the method of choice. The surgical character of the experiments was evident. It could be seen, for example, in Bernard's invention of a new operative procedure for sectioning the fifth pair of nerves inside the skull, or in his use of surgical instruments (cataract needle, stylet) in his studies of reflex phenomena and of the sympathetic nerve.[43]

Coordination of experimental results with clinical and pathological findings persisted. For example, Bernard noted similarities between the rotatory phenomena and presence of sugar and albumin in the urine of rabbits whose middle cerebellar peduncles had been sectioned and the albuminuria observed in eclamptics.[44] In another, more famous case Bernard compared the production of excess sugar in the urine resulting from experimental wounding of the floor of the fourth ventricle in rabbits with diabetes.[45] Yet perhaps because Bernard no longer practiced medicine and was consolidating his vocation as a physiologist, such correlations occupied a less prominent place in his arguments after 1844. The strong interaction of laboratory and clinical findings that was so marked in Magendie's studies on the nervous system gave way to a situation in which physiology informed pathology, while the latter provided at best ancillary support for the former.

That the balance was not tipped still further away from medical concerns and perspectives was due in part to the continuing interest and cooperation of a number of Paris physicians. Foremost among these was Bernard's old patron and the first president of the Société de Biologie, Pierre Rayer. Rayer was convinced of the value of clinical and pathological experience for physiology, and placed at Bernard's disposal observations and material from his own wards at the Charité. On one occasion, a clinical case of Rayer's was cited to support the assertion that motor function depended on the integrity of the sensitive roots of the spinal cord. In another case, Bernard performed an autopsy on a diabetic who had died on Rayer's service, noted the quantity of sugar present at various locations, and compared the results to other autopsies and to the results of animal experiments. Rayer's scientific involvement was more directly reflected in the meticulous dissection of a two-headed calf, on which he collaborated with Bernard, and which concluded with joint reflections on the physiological implications of the findings.[46] Indeed, Rayer was an active partisan of the pathological physiology defined by Magendie and Scipion Pinel in the 1820s, and he carried that program into the life of the Société de Biologie in the 1850s.

The mark of its medical context remained on Bernard's physiology in still another sense: its almost exclusive concentration on the elucidation of human functions. The comparative dimension, which was to expand into a genuinely biological enterprise in Bernard's later writings, was still severely delimited. Indeed, it even appeared to contract in the period 1845–1853. In the midst of the familiar dogs, cats, rabbits, and horses, even the occasional frog seemed an exotic intruder among the experimental animals.

Some studies were completely devoid of comparative observations.[47]

The physiological interest in drugs continued and expanded. In part, it took the form of an extension and a refinement of Magendie's investigations on drug action, as in Bernard's attempt to demonstrate that the first action of strychnine was not on the spinal cord but on the peripheral extremities of the sensitive nerves.[48] He promised studies of other such substances having obvious action on the nervous system, such as hydrocyanic acid, opium, and nicotine. In 1849 he reported to the Société de Biologie the results of experiments comparing the effects of atropine on dogs and rabbits.[49] In part, the interest was expressed in the use of drugs as analytical tools, to differentiate the effects of specific nerves, or of procedures that affected nerves. Such was his use of the recently discovered anesthetics ether and chloroform in studies of recurrent sensitivity and of the effects of sectioning the sympathetic nerve on body heat. The same impulse motivated his use of the vegetable poison curare in studies of the effects of the sympathetic nerve on sensibility.[50]

At least one theme became more explicit, and more sharply defined, than it had been for Magendie. This was the resolution of apparent contradictions in physiological results by the precise specification of experimental conditions. The case in point usually cited is Bernard's determination of conditions necessary for the expression of recurrent sensitivity. Another case, equally elegant though less well known, is Bernard's work on the turning motion following lesion of the middle cerebellar peduncles. Magendie had found that wounding this organ resulted in violent movements of rotation in the direction of the wounded side. G. V. Lafargue and Longet, working independently, had obtained opposite results. Bernard found that the direction of rotation depended on whether the peduncle was wounded behind or in front of the origin of the nerve of the fifth cranial pair. In the former case the animal would turn in the direction of the side wounded, in the latter case to the opposite side.[51] Later, in a different context, Bernard drew the moral by remarking that "when, in physiology, a phenomenon presents itself with contradictory appearance, we may be assured that its elements are still complex and that its conditions of existence have not yet been sufficiently analyzed."[52]

By 1853 Bernard had begun to move away from the naive

empiricist posture of his early memoirs. Reflecting on the failure of earlier investigators, himself included, to perceive the connection between sectioning the sympathetic nerve and increased heat, circulation, and sensitivity in the parts supplied by the nerve, he was compelled to recognize the importance of a prepared mind. "The experiments," he remarked, "had to give the same results in the hands of all experimenters, who consequently had to have the phenomenon in question under their eyes. But it is easy to have a phenomenon under one's eyes and not to see it, so long as some circumstance does not direct the mind toward it." Between 1842 and 1852, Bernard continued, he had been "placed in a different point of view for observing the results of the experiment."[53] What had produced that new point of view—his joint studies with Magendie on animal heat, his own intervening work on other aspects of the nervous system, or some other factor—Bernard does not say. And he does not yet speak in terms of the utility and necessity of hypotheses or preconceived ideas. But he was by now aware of the controlling role in observation played by the mental state of the investigator.

Juxtaposition of Magendie's later work on the nervous system with Bernard's research in the same area over the first ten or twelve years of his scientific career shows that at least until 1853 Bernard did not diverge significantly from the patterns set by Magendie. He did not move away from a focus on human physiology.

Nor had the Claude Bernard of these years elaborated a fully coherent methodological position. By 1853 he was just beginning to move away from a naive empiricist posture and toward an appreciation of the role of ideas in research. In this he was only retracing the steps already taken by Magendie more than ten years earlier, as the latter had reflected on the discovery and reception of the cerebrospinal fluid. Though incipient tendencies to generalization may be detected in his early work, Bernard had not yet propounded any of the great physiological principles—such as the notion of the internal environment—which he was to seek with so much zeal in his later career.[54]

In his concrete investigations on the nervous system, he continued to rely on Magendie's methods of choice: animal experiment of a marked surgical quality, and clinical and pathological observation. His choice of research problems was sometimes original, but again—as in the cases of the chorda tympani and spinal nerves, recurrent sensitivity, or the action of strychnine on the nervous system—they were just as often taken over from Magendie.

If there was a divergence worth noting, it lay in part in Bernard's superior versatility and ingenuity in experimental design, and his more consistent and self-aware pursuit of an unambiguous outcome in experimental investigations. In part, it was expressed in Bernard's greater emphasis on the leading role of animal experiment vis-à-vis pathological experience or comparative observation, a shift clearly associated with his move away from medical practice. The same emphasis was reflected in the official program of the new Société de Biologie, in which Bernard shared vice-presidential responsibilities with Charles-Phillipe Robin.

Robin was a product of the Paris Faculty of Medicine who, like Bernard, found science more to his taste than medical practice. A year after taking his M.D. in 1846 he presented two theses on biological topics for his *docteur ès sciences,* and in 1849 he was named to the chair of natural history at the Faculty of Medicine. Through Bernard's patron and collaborator Pierre Rayer he was introduced to Auguste Comte, and the latter's lectures won him over to the ideal of a general science of biology.[55]

It was Robin who set out the goals of the Société in the first volume of its *Comptes rendus et mémoires.* Most emphatically, pure science was to be given pride of place. According to Robin it was through the medical art that knowledge of physiology, anatomy, and pathology first arose; but once constituted as sciences, those subjects could best be pursued apart from application. No doubt the applications would come, just as they had for physics and chemistry. But they had to be regarded as derivative of the abstract or pure sciences. Similarly, Robin argued that pathology had to be subordinated to physiology. Pathology should start from knowledge of the normal state and make known its alterations. The experience of clinic and hospital, even when edited and rationalized, would not suffice: "Our society is not a society of pathological anatomy nor of pathology." All of the phenomena of life, plant as well as animal, would be embraced in the Société's field of study.[56]

Clearly the founders of the Société de Biologie, speaking through Robin, wished to distance themselves from what they perceived as the excessively narrow and practical focus of the mainstream of the Paris clinical school. They would constitute an autonomous science of life, or biology, much along the lines envisioned by Auguste Comte.[57] Yet for all that, even their formal statement of purpose bore the deep imprint of the medical milieu in which it was formulated. Advancement of the medical art remained the ultimate, if not the proximate goal: "For us it is the most important—that toward which all our works naturally tend." Consequently, human structure and function remained of paramount interest. Comparative studies, Robin urged, should always be pursued with an eye to the illumination of the human organism. Comparative anatomy—surprisingly—was not even recognized as a distinct science but was considered to be one among several methods of biological investigation. Clinical experience and pathological anatomy, though not objects of study in themselves, would be utilized as much as possible "as so many completed experiments" that might clarify normal structure and function.[58]

The medical imprint is still clearer if we turn from this statement of intent to the realities of the Société's composition and behavior. Of the Société's members, the overwhelming majority of those most directly involved in its activities were M.D.'s or had some close connection with the healing arts.[59] The Société's first president, Pierre Rayer, was a leading Paris clinician; and Robin, one of its first vice-presidents, was a professor at the Faculty of Medicine. Weekly meetings were held in Robin's amphitheater at the Ecole Pratique.[60] At least in its early years, the Société's official publication was dominated by the sciences closest to medicine, with pathological subjects claiming the most entries.[61]

Neither in ambition nor in substance did the Société de Biologie embody a sharp break with the past. Much of its program and more of its practice were already present in Magendie's *Journal* of the 1820s, and in the research of Magendie and others in the intervening decades. In particular, Magendie's ideal of pathological physiology was preserved.

What is visible, at least in the Société's stated program, is an unmistakable shift in the balance between physiology and pathology, and between pure science and its applications in the medical art. Physiology would lead; pathology would follow. Science would come first; its informing of medical practice later, perhaps much later. Magendie, always as much physician as physiologist, could never so subordinate the pathological to the normal, nor wait with such equanimity for the improvement of practice. That Bernard could do so, and that other members of the Société de Biologie could strive to do so, reflected a significant change of attitude within the more scientifically minded segment of the Paris medical community. The era of the 1820s and 1830s, when clinical medicine and physician-generated pathological observation were vigorously ascendent, had passed. The era of the laboratory, of the experimental ideal, and of the specialist in research lay ahead. Magendie, who for so long had achieved a remarkable equilibrium in his roles as physician and scientist, now made way for Bernard, the first of a new species, the research physiologist. In the decades after 1850 Bernard would move decisively beyond his teacher, both in his extension of the applications of physiology to pathology and in his broadening of human into general physiology. Without abandoning its ambition to create a scientific foundation for medicine, physiology could begin to take its place as a field of biology, in the vanguard of that science's experimental ideal.

Notes

Index

Abbreviations

AC	*Annales de chimie*
ACP	*Annales de chimie et de physique*
BU	*Biographie universelle*
BHM	*Bulletin of the History of Medicine*
DSB	*Dictionary of Scientific Biography*
HMSRM	*Histoire et mémoires de la Société Royale de Médecine*
JHB	*Journal of the History of Biology*
JHMAS	*Journal of the History of Medicine and Allied Sciences*
JPEP	*Journal de physiologie expérimentale et pathologique*
SHB	*Studies in History of Biology*

Notes

Introduction

1. On *Naturphilosophie* see Karl E. Rothschuh, *History of Physiology* (Huntington, N.Y.: Krieger, 1973), pp. 155–166, 172. For furthur references consult Max Neuburger, *The Historical Development of Experimental Brain and Spinal Cord Physiology before Flourens*, trans. and ed. Edwin Clarke (Baltimore: Johns Hopkins University Press, 1981), p. 191, n. 21. On teleomechanism see Timothy Lenoir, *The Strategy of Life: Teleology and Mechanics in Nineteenth Century German Biology* (Dordrecht, Holland: D. Reidel, 1982), esp. pp. 68–71, 103–106. On German medical education see Edwin Lee, *Observations on the Principal Medical Institutions and Practice of France, Italy, and Germany* (London, 1843), pp. 166–205; Theodor Billroth, *The Medical Sciences in the German Universities* (New York: Macmillan, 1924), p. 92; and Hans H. Simmer, "Principles and Problems of Medical Undergraduate Education in Germany During the Nineteenth and Early Twentieth Centuries" in *The History of Medical Education*, ed. C. D. O'Malley (Berkeley: University of California Press, 1970), pp. 173–200.

2. Gerald L. Geison, *Michael Foster and the Cambridge School of Physiology* (Princeton: Princeton University Press, 1978), pp. 13–47; and June

Goodfield-Toulmin, "Some Aspects of English Physiology: 1740–1840," *JHB*, 2 (1969): 283–320. On antivivisection see J. M. D. Olmsted, *François Magendie, Pioneer in Experimental Physiology and Scientific Medicine in Nineteenth Century France* (New York: Schuman, 1944), pp. 116–117, 137–143; Geison, *Michael Foster*, pp. 18–21; Neuburger, *Experimental Brain and Spinal Cord Physiology*, pp. 190–191, 213–214; *Letters of Sir Charles Bell, Selected from His Correspondence with His Brother George Joseph Bell* (London, 1870), pp. 161, 275–276; H. Bretschneider, *Der Streit um die Vivisektion im 19. Jahrhundert* (Stuttgart: G. Fischer, 1962); and Richard D. French, *Antivivisection and Medical Science in Victorian Society* (Princeton: Princeton University Press, 1975).

3. Edward C. Atwater, " 'Squeezing Mother Nature': Experimental Physiology in the United States Before 1870," *BHM*, 52 (1978): 313–335.

1. A Science in the Making

Epigraphs: Philippe Pinel and Pierre-François Percy, in Institut de France, Académie des Sciences, *Procès verbaux*, 5 (1812–1815): 205–208 (meeting of May 3, 1813). Etienne Geoffroy Saint-Hilaire, "1° La zoologie a-t-elle, dans l'Académie des Sciences, une représentation suffisante?— 2° La physiologie n'y a-t-elle pas été entièrement oubliée?" *Revue encyclopédique ou analyses et announces raisonnées des productions les plus remarquables dans la littérature, les sciences et les arts*, 13 (1822): 501–511.

1. Institut de France, Académie des Sciences, *Procès verbaux*, 1 (An IV–VII/1795–1799):1.

2. The sections were Medicine and Surgery (IX) and Rural Economy and the Veterinary Art (X). Institut de France, Académie des Sciences, *Procès verbaux*, 1 (An IV–VII/1795–1799):1.

3. *Oxford English Dictionary*, vol. 7 (Oxford: Clarendon Press, 1933), p. 811. Fontenelle's history of the Paris Academy of Science for 1666–1699 does not use the term "physiology." On this point, and on the similar broad meaning of "physics" through the early eighteenth century, see Claire Salomon-Bayet, *L'institution de la science et l'expérience du vivant: méthode et expérience à l'Académie Royale des Sciences, 1666–1793* (Paris: Flammarion, 1978), pp. 43–54, 126–127, 370–371; and J. L. Heilbron, *Electricity in the 17th and 18th Centuries: A Study of Early Modern Physics* (Berkeley: University of California Press, 1979), pp. 9–13.

4. Georges Canguilhem, "La constitution de la physiologie comme science," in *Etudes d'histoire et de philosophie des sciences*, 3rd ed. (Paris: Vrin, 1975), pp. 226–273, esp. p. 226.

5. Karl Rothschuh, *History of Physiology* (Huntington, N.Y.: Krieger, 1973), p. 120.

6. A. Corlieu, *L'enseignement au Collège de Chirurgie depuis son origine jusqu'à la révolution française* (Paris, 1890), pp. 12–20 (the chair of Phys-

iology was later renamed Physiology and Hygiene); Albrecht Haller, *Elémens de physiologie,* traduction nouvelle du Latin en Français, par M. Bordenave (Paris, 1769).

7. Rothschuh, *History of Physiology,* p. 154; Guenter B. Risse, "Reil, Johann Christian," *DSB,* 11 (1975): 363–365.

8. A. Rupert Hall, *The Scientific Revolution, 1500–1800,* 2nd ed. (Boston: Beacon, 1972), pp. 152–153, 288–289; Robert G. Frank, *Harvey and the Oxford Physiologists* (Berkeley: University of California Press, 1980).

9. See G. A. Lindeboom, "Sylvius, Franciscus de le Boë," *DSB,* 13 (1976): 222–223; E. Ashworth Underwood, "Franciscus Sylvius and His Iatrochemical School," *Endeavor,* 31 (1972): 73–76; Walter Pagel, "Helmont, Johannes (Joan) Baptista van," *DSB,* 6 (1972): 253–259; R. M. Rattansi, "The Helmont-Galenist Controversy in Restoration England," *Ambix,* 12 (1964): 1–23; Thomas B. Settle, "Borelli, Giovanni Alfonso," *DSB,* 2 (1970): 306–314; M. D. Grmek, "Baglivi, Georgius," *DSB,* 1 (1970): 391–392; Lester S. King, *The Road to Medical Enlightenment, 1650–1695* (New York: American Elsevier, 1970), 37–62, 88–90, 93–112; Allen G. Debus, *The Chemical Philosophy,* vol. 2 (New York: Science History Publications, 1977), pp. 295–379, 502–537; idem, *The English Paracelsians* (London: Oldbourne, 1965), pp. 181–183; E. Bastholm, *History of Muscle Physiology,* Acta Historica Scientarum Naturalium et Medicinalium, vol. 7 (Copenhagen: Munksgaard, 1950), pp. 163–189; and Theodore Brown, *The Mechanical Philosophy and the Animal Oeconomy* (New York: Arno, 1981).

10. On Trembley, see John R. Baker, *Abraham Trembley of Geneva: Scientist and Philosopher, 1710–1784* (London: Edward Arnold, 1952); on Humboldt, Kurt Biermann, "Humboldt, Friedrich Wilhelm Heinrich Alexander von," *DSB,* 6 (1972): 549–555; on Fontana, Luigi Belloni, "Fontana, Felice," *DSB,* 5 (1972): 55–57; M. P. Earles, "The Experimental Investigation of Viper Venom by Felice Fontana (1730–1805)," *Annals of Science,* 16 (1960): 255–268; and Peter K. Knoefel, *Felice Fontana, 1730–1805: An Annotated Bibliography* (Trent: Temi, 1980); on Spallanzani, Jean Rostand, *Les origines de la biologie expérimentale et l'abbé Spallanzani* (Paris: Fasquelle, 1951); on Hales, Henry Guerlac, "Hales, Stephen," *DSB,* 6 (1972): 35–48; and on Réaumur, J. B. Gough, "Réaumur, René Antoine Ferchault de," *DSB,* 11 (1975): 327–335.

11. Rothschuh, *History of Physiology,* pp. 114–115; Guenter B. Risse, "Hoffmann, Friedrich," *DSB,* 6 (1972): 458–461.

12. L. J. Rather, "G. E. Stahl's Psychological Physiology," *BHM,* 35 (1961): 37–49; Jacques Roger, *Les Sciences de la vie dans la pensée française du XVIIIᵉ siècle* (Paris: A. Colin, 1963): 427–431; and Max Neuburger, *The Historical Development of Experimental Brain and Spinal Cord Physiology Before Flourens,* trans. and ed. Edwin Clarke (Baltimore: Johns Hopkins University Press, 1981), esp. pp. 113–115, 153–168.

13. Bernward Josef Gottlieb, "Bedeutung und Auswirkungen des hallischen Professors und Kgl. preuss. Leibarztes Georg Ernst Stahl auf

den Vitalismus des XVIII Jahrhunderts, insbesondere auf die Schule von Montpellier," *Nova Acta Leopoldina*, n.s., 12 (1943): 425–502.

14. Cf. Lester S. King, *The Medical World of the Eighteenth Century* (Huntington, N.Y.: Krieger, 1971; rpt. of 1958 ed.), pp. 59–121; Rothschuh, *History of Physiology*, pp. 115–119; G. A. Lindeboom, *Herman Boerhaave: The Man and His Work* (London: Methuen, 1968), esp. pp. 70–75, 264–282; idem, ed., *Boerhaave and His Time* (Leiden: Brill, 1970).

15. For Haller's biography see Erich Hintzsche, "Haller, (Victor) Albrecht von," *DSB*, 6 (1972): 61–67.

16. Albrecht Haller, "A Dissertation on the Sensible and Irritable Parts of Animals," rpt. of 1755 English trans. with introd. by Owsei Temkin, *BHM*, 4 (1936): 651–699; idem, *Deux mémoires sur le mouvement du sang, et sur les effets de la saignée, fondés sur des expériences faites sur des animaux* (Lausanne, 1756); idem, "Mémoire sur la formation des os," in Fougeroux, ed., *Mémoires sur les os* (Paris, 1760); and Neuburger, *Experimental Brain and Spinal Cord Physiology*, esp. pp. 113–152. In the introduction to the first volume of his *Elementa physiologiae corporis humani* (Lausanne, 1756–1766) Haller made a strong statement on behalf of animal experiment. Cf. Franz Heineman, *Albrecht von Haller als Vivisektor* (Bern, 1908).

17. Albrecht Haller, *First Lines of Physiology*, trans. from 3rd Latin ed. (Troy, N.Y.: Penniman, 1803).

18. See, for example, ibid., 18–19, 94–99, 181–183.

19. Erich Hintzsche and Jörn Henning Wolf, *Albrecht von Hallers Abhandlung über die Wirkung des Opiums auf den menschlichen Körper* (Bern: Haupt, 1962). The diversity of Haller's research approach is sometimes obscured in the interest of setting off the eighteenth century neatly against the nineteenth. Cf. Joseph Schiller, "Physiology's Struggle for Independence in the First Half of the Nineteenth Century," *History of Science*, 7 (1968): 64–89, esp. p. 73.

20. Jacques Lordat, *Exposition de la doctrine médicale de P. J. Barthez et mémoires sur la vie de ce médecin* (Paris, 1818), pp. 58–59; and Neuburger, *Experimental Brain and Spinal Cord Physiology*, pp. 113–152.

21. Rothschuh, *History of Physiology*, pp. 81, 103, 117, 119, 139–140, 143–144; David R. Dyck, "Eller von Brockhausen, Johann Theodor," *DSB*, 4 (1971): 352–353; Hans Straub, "Bernoulli, Daniel," *DSB*, 2 (1970): 36–46; Aubrey B. Davis, "Lieberkühn, Johannes Nathanael," *DSB*, 8 (1973): 327–328; Vladislav Kruta, "Prochaska, Georgius (Jiri)," *DSB*, 11 (1975): 158–160; David M. Knight, "Girtanner, Christoph," *DSB*, 5 (1972): 411.

22. William Coleman, "Kielmeyer, Carl Friedrich," *DSB*, 7 (1973): 366–369; and Timothy Lenoir, *The Strategy of Life: Teleology and Mechanics in Nineteenth Century German Biology* (Dordrecht, Holland: D. Reidel, 1982), esp. pp. 37–53.

23. Rothschuh, *History of Physiology*, pp. 140–143; Erich Hintzsche, "Caldani, Leopoldo Marcantonio" *DSB*, 3 (1971): 15–16; Dorothy M. Schullian, "Cotugno, Domenico Felice Antonio," *DSB*, 3 (1971): 437–

438; Theodore M. Brown, "Galvani, Luigi," *DSB*, 5 (1972): 267–269; and Neuburger, *Experimental Brain and Spinal Cord Physiology*, esp. pp. 139–143.

24. Rothschuh, *History of Physiology*, 132–133, 137, 140, 143; Luigi Belloni, "Fontana, Felice," *DSB*, 5 (1972): 55–57; Claude E. Dolman, "Spallanzani, Lazzaro," *DSB*, 12 (1975): 553–567; Rostand, *Les origines*; and Knoefel, *Felice Fontana*.

25. Rothschuh, *History of Physiology*, pp. 131–143; Theodore M. Brown, "Stuart, Alexander," *DSB*, 13 (1976): 121–123; "Jurin, James," *Dictionary of National Biography*, 10 (London: Smith, Elder, 1908): 1117–18; William LeFanu, "Hewson, William," *DSB*, 6 (1972): 367–368; Jessie Dobson, "Hunter, John," *DSB*, 6 (1972): 566–568; Stephen J. Cross, "John Hunter, the Animal Oeconomy, and Late Eighteenth-Century Physiological Discourse," *SHB*, 5 (1981): 1–110; Jessie Dobson, "Cruikshank, William," *DSB*, 3 (1971): 486–488; Samuel X. Radbill, "Whytt, Robert," *DSB*, 14 (1976): 319–324; William P. D. Wightman, "Cullen, William," *DSB*, 3 (1971): 494–495; and Henry Guerlac, "Hales, Stephen."

26. On the role of the Scottish universities in the nineteenth century, see Gerald L. Geison, *Michael Foster and the Cambridge School of Physiology* (Princeton: Princeton University Press, 1979), pp. 43–44.

27. Brown, "Stuart, Alexander."

28. Gough, "Réaumur."

29. Corlieu, *L'enseignement au Collège de Chirurgie*, pp. 12–20; Haller, *Elémens de physiologie*. See note 9, above. The name was probably taken over, with the textbook, from Haller.

30. A. E. Best, "Pourfour du Petit, François," *DSB*, 11 (1975): 111–113; Pierre Huard and Marie José Imbault-Huart, "Lorry, Anne Charles," *DSB*, 8 (1973): 505–507; "Saucerotte, Louis Sébastien," *BU*, 38 (Paris, n.d.): 42–43; "Petit, Jean-Louis," *BU*, 32 (Paris, n.d.): 594–596; "Tenon, Jacques-René," *BU*, 41 (Paris, n.d.): 147–148; Neuburger, *Experimental Brain and Spinal Cord Physiology*, pp. 68–70, 90–100, 171–175. Lorry, an M.D., taught surgery at the Paris Faculty of Medicine.

31. Antoine Louis, *Lettres sur la certitude des signes de la mort* (Paris, 1752), pp. 223–250; *Mémoire sur une question anatomique relative à la jurisprudence* (Paris, 1763); *Rapport des expériences faites par l'Académie Royale de Chirurgie, sur différentes méthodes de tailler* (n.p., n.d.); Toby Gelfand, "The Training of Surgeons in Eighteenth-Century Paris and Its Influence on Medical Education" (diss., Johns Hopkins University, 1973), pp. 156–157. On Louis see Pierre Huard and Marie-José Imbault-Huart, "Antoine Louis (1723–1792), Secrétaire perpétuel de l'Académie Royale de Chirurgie," in *Biographies médicales et scientifiques*, ed. Pierre Huard (Paris: Roger Dacosta, 1972), pp. 33–116.

32. Louis Dulieu, "Le mouvement scientifique montpellieran au XVIIIᵉ siècle," *Revue d'histoire des sciences*, 11 (1958): 227–249. There was no chair designated specifically for physiology.

33. On Sauvages see F. Bérard, *Doctrine médicale de l'école de Mont-*

pellier, et comparaison de ces principes avec ceux des autres écoles d'Europe (Montpellier, 1819), pp. 44—52; Louis Dulieu, "Le mouvement," pp. 238, 241—242; idem, "François Boissier de Sauvages (1706—1767)," *Revue d'histoire des sciences*, 22 (1969): 303—322.

34. Théophile Bordeu, *Oeuvres complètes* (Paris, 1818); Thomas S. Hall, *Ideas of Life and Matter*, vol. 2 (Chicago: University of Chicago Press, 1969), pp. 82—86; *BU*, 5 (Paris, 1843): 71—74.

35. P. J. Barthez, *Nouveaux élémens de la science de l'homme* (Montpellier, 1778); Lordat, *Barthez*, esp. pp. 2—33, 42—44, 49—73, 109—117, 125—126, 132—148, 198—201, 217—218, 247; Dulieu, "Le mouvement," pp. 237—239; Bérard, *Montpellier*, pp. 97—98.

36. Lordat, *Barthez*, pp. 110—113, 263—271; John Cross, *Paris et Montpellier, ou tableau de la médecine dans ces deux écoles*, trans. from the English by Elie Revel (Paris and Montpellier, 1820), pp. 177—187; Hall, *Ideas*, vol. 2, pp. 87—91. In contrast to the practice of later historians, the application of the term "vitalism" around 1800 appears to have been restricted to the doctrine of Barthez and its derivatives. Haller, Bordeu, and Bichat are termed "solidists." See Cross, *Paris et Montpellier*, pp. 44—47, 176—187.

37. Pierre-Jean-Georges Cabanis, *Du degré de certitude de la médecine* (Paris, 1804), in *Oeuvres philosophiques de Cabanis*, ed. Claude Lehec and Jean Cazeneuve, pt. 1 (Paris: Presses Universitaires de France, 1956), pp. 33—103, esp. pp. 57—62, 87. Notes indicate all changes from the 1798 edition.

38. Bernard himself might have written Barthez's second rule of method from the *Discours préliminaire* of the *Nouveaux élémens*: "All that we can learn concerning the cause of a phenomenon reduces to ascertaining by exact observation the circumstances, laws, and conditions of its production." Quoted in Lordat, *Barthez*, pp. 117—120.

39. Salomon-Bayet, *L'institution*, esp. pp. 109—113, 179—181, 211; James E. McClellan III, "The Académie Royale des Sciences, 1699—1793: A Statistical Portrait," *Isis*, 72 (1981): 541—567.

40. Yves Laissus, "Le Jardin du Roi," in *Enseignement et diffusion des sciences en France au XVIIIᵉ siècle*, ed. René Taton (Paris: Hermann, 1964), pp. 287—341, esp. 313—314; Charles Coury, *L'enseignement de la médecine en France des origines à nos jours* (Paris: Expansion Scientifique Française, 1968), p. 17; Charles Coulston Gillispie, *Science and Polity in France at the End of the Old Regime* (Princeton: Princeton University Press, 1980), pp. 130—184.

41. Vern Bullough, *The Development of Medicine as a Profession: The Contribution of the Medieval University to Modern Medicine* (New York: Hafner, 1966); Gelfand, "Training of Surgeons," pp. 1—16.

42. The phases of this process included the merger of surgeons and barber-surgeons guilds in the mid-seventeenth century; the issuance of a royal edict in 1692 making a definitive separation of surgeons from barber-wigmakers; the foundation of a royally supported Collège de Chirurgie in 1724 and of an Académie de Chirurgie in 1731; the im-

position of the requirement of master of arts on future surgeons in Paris from 1743; and the establishment of the doctorate in surgery, including a Latin thesis, from 1749. See Toby Gelfand, *Professionalizing Modern Medicine: Paris Surgeons and Medical Science and Institutions in the 18th Century* (Westport, Conn.: Greenwood Press, 1980), pp. 21–79; Pierre Huard, L'instruction médico-chirurgical" in Taton, *Enseignement*, pp. 191–192; and Marie-José Imbault-Huart, "Les chirurgiens et l'esprit chirurgical en France au XVIIIème siècle," *Clio medica*, 15 (1981): 143–157.

43. The original five chairs created at the Collège in 1724 included one of human anatomy, and in 1755 the original chair of principles of surgery was renamed the chair of physiology. In the same year the Collège added a chair of surgical chemistry, and in 1791 a chair of botany. Corlieu, *L'enseignement au Collège de Chirurgie*, pp. 12–20.

44. Charles Bedel, "L'enseignement des sciences pharmaceutiques," in Taton, *Enseignement*, pp. 237–257; Gillispie, *Science and Polity*, pp. 203–212.

45. Pierre Huard, "L'instruction médico-chirurgical," pp. 206–209; Clément Bressou, *Histoire de la médecine vétérinaire* (Paris: Presses Universitaires de France, 1970), pp. 64–76; Caroline C. Hannaway, "Veterinary Medicine and Rural Health Care in Pre-Revolutionary France," *BHM*, 51 (1977): 431–437.

46. Coury, *L'enseignement de la médecine*, pp. 18, 99–101; Pierre Huard, "L'instruction médico-chirurgical," p. 210.

2. Context for Change

Epigraph: Jacques Tenon, "Reflexions sur cette question: Enseignera-t-on séparément l'anatomie et la physiologie dans les nouvelles ecoles de médecine?" (c. 1794), fol. 169–172, *Nouvelles acquisitions françaises*, no. 22749, Bibliothèque Nationale, Paris.

1. Pierre-Jean-George Cabanis, *Du degré de certitude de la médecine* (Paris, 1804), in *Oeuvres philosophiques de Cabanis*, ed. Claude Lehec and Jean Cazeneuve, pt. 1 (Paris: Presses Universitaires de France, 1956), p. 41; notes indicate all changes from the 1798 edition. For the date of completion of this essay see François Joseph Picavet, *Les Idéologues: Essai sur l'histoire des idées et des théories scientifiques, philosophiques, religieuses etc. en France depuis 1789* (Paris, 1891), p. 184. For Cabanis's biography see Lehec and Cazeneuve, eds., *Oeuvres*, pp. v–xxi; Georges Canguilhem, "Cabanis, Pierre-Jean-George," *DSB*, 3 (1971): 1–3; and Martin S. Staum, *Cabanis: Enlightenment and Medical Philosophy in the French Revolution* (Princeton: Princeton University Press, 1980), pp. 14–19, 94–99. Staum's book is now the standard work on Cabanis.

2. Sergio Moravia, "Philosophie et médecine en France à là fin du XVIIIe siècle," *Studies on Voltaire and the Eighteenth Century*, 89 (1972): 1089–1151; Erna Lesky, "Cabanis und die Gewissheit der Heilkunde," *Gesnerus*, 11 (1954): 152–182; Guenter B. Risse, "The Quest for Cer-

tainty in Medicine: John Brown's System of Medicine in France," *BHM*, 45 (1971): 1–12.

3. For Rousseau see the *Oeuvres complètes*, ed. P. R. Auguis (Paris, 1824), vol. 3, *Emile: Livre premier*, pp. 51–56. The unfavorable contrast of medicine with the physical sciences has been underlined in Harold J. Abrahams, "Lavoisier's Proposals for Training in Science and Medicine," *BHM*, 32 (1958): 389–407. The vogue for mesmerism in the 1780s owed something to popular distrust of orthodox medicine. See Robert Darnton, *Mesmerism and the End of the Enlightenment in France* (Cambridge, Mass.: Harvard University Press, 1968), p. 15.

4. Cabanis, *Oeuvres*, vol. 1, pp. 57–62. On this point and on the essay as a whole see Staum, *Cabanis*, pp. 103–109.

5. Cabanis, *Oeuvres*, vol. 1, pp. 64–73.

6. Ibid, pp. 74–79, 80–93. Cabanis's empiricist stance again echoes that of Barthez. Cabanis's debts to the Montpellier school are explored in more detail in Martin S. Staum, "Medical Components in Cabanis's Science of Man," *SHB*, 2 (1978): 1–31; and idem, *Cabanis*, pp, 78–90.

7. Marc-Antoine Petit, "Eloge de Pierre-Joseph Desault, chirurgien en chef de l'Hôtel-Dieu de Paris, prononcé à l'ouverture des cours d'anatomie et de chirurgie de l'Hôtel-Dieu de Lyon, le 5 décembre 1795," MS 5212, Bibliothèque de la Faculté de Médecine, Paris; Toby Gelfand, *Professionalizing Modern Medicine: Paris Surgeons and Medical Science and Institutions in the 18th Century* (Westport, Conn.: Greenwood Press, 1980), pp. 183–184.

8. *HMSRM, 1787–1788* (Paris, 1790), p. 16.

9. Gelfand, *Professionalizing Modern Medicine*, p. 184.

10. Antoine Fourcroy, *Rapport à la Convention au nom des Comités de Salut Public et d'Instruction Publique* (Paris, An III/1795), p. 9.

11. Tenon, "Reflexions"; Gelfand, *Professionalizing Modern Medicine*, pp. 184–185.

12. Tenon, "Reflexions"; Jacques Tenon, "Sur l'exfoliation des os," in Académie Royale des Sciences, *Mémoires, 1758* (Paris, 1763), pp. 372–418, and *Mémoires, 1760* (Paris, 1766), pp. 223–238.

13. For a discussion of parallels in the situation of medicine in contemporary German-speaking lands, see Lesky, "Cabanis."

14. Joseph Fayet, *La révolution française et la science, 1789–1795* (Paris: M. Rivière, 1960), pp. 203–204; Roger Hahn, *The Anatomy of a Scientific Institution: The Paris Academy of Sciences, 1666–1803* (Berkeley: University of California Press, 1971), pp. 278–280; R. R. Palmer, *Twelve Who Ruled* (Princeton: Princeton University Press, 1970; orig. pub. 1941), pp. 127–128.

15. Abrahams, "Lavoisier's Proposals," p. 396; Hahn, *Anatomy*, pp. 284–285; Pierre Huard and H. D. Grmek, *Sciences, médecine, pharmacie de la Révolution à l'Empire, 1789–1815* (Paris: Roger Dacosta, 1970), pp. 120–121; Lesky, "Cabanis," pp. 172–173; Louis Liard, *L'enseignement supérieur en France, 1789–1889* (Paris, 1888), pp. 276–277; Gelfand,

Professionalizing Modern Medicine, p. 163; Paul Triaire, *Récamier et ses contemporains, 1774–1852* (Paris, 1899), p. 88.

16. Liard, *L'enseignement*, p. 277; J. M. D. Olmsted, "French Medical Education as a Legacy from the Revolution," in *Essays in Biology in Honor of Herbert Evans* (Berkeley: University of California Press, 1943), pp. 464–466.

17. Olmsted, "French Medical Education," pp. 464–466. For discussion of the background of the unification of medicine and surgery, see Gelfand, *Professionalizing Modern Medicine*, pp. 149–171.

18. A. Corlieu, *Centennaire de la Faculté de Médecine de Paris, 1794–1894* (Paris, 1896), pp. 247–248, 262; Huard and Grmek, *Sciences, médecine, pharmacie*, pp. 123, 144; Gelfand, *Professionalizing Modern Medicine*, pp. 90–92. Government regulation of medical practice followed a different course. See Toby Gelfand, "The Training of Surgeons in Eighteenth-Century Paris and Its Influence on Medical Education" (diss., Johns Hopkins University, 1973), pp. 422–427; and Olmsted, "French Medical Education," pp. 467–468.

19. Louis Liard, *L'Université de Paris* (Paris: Librairie Renouard–H. Laurens, 1909), pp. 30–34.

20. The same year witnessed the first publication of the *Journal de la Société des Pharmaciens de Paris*, which continued until 1799. Following a hiatus of ten years, during which pharmacists published in the *Annales de chimie*, their specialist journal reappeared in 1809 under the name of *Bulletin de pharmacie et des sciences accessoires*. See L. André-Pontier, *Histoire de la pharmacie* (Paris: O. Doin, 1900), pp. 252–253; René Fabre and Georges Dilleman, *Histoire de la pharmacie* (Paris: Presses Universitaires de France, 1963), pp. 36–37; and Liard, *L'Université de Paris*, pp. 30–34.

21. The *Nouveau plan* had also called for the maintenance within the hospital of a group of experimental animals on which drugs prepared by the pharmacist could be employed before their use by the physician. This proposal—which in its association of animal experiment and chemical pharmaceutical knowledge in the study of drug action immediately calls to mind the later research programs of Magendie and Bernard—advanced ideas identical to those of the veterinarian Claude Bourgelat. Bourgelat had recommended that the doors of the veterinary schools always be open to those wishing to make experiments in comparative medicine, and had been convinced that the study of animal and human illness could be mutually illuminating. Huard and Grmek, *Sciences, médecine, pharmacie*, pp. 157–160; André-Pontier, *Histoire*, pp. 245–248.

22. Dora B. Weiner, "Public Health under Napoleon: The Conseil de Salubrité de Paris, 1802–1815," *Clio medica*, 9 (1974): 271–284; Ann Fowler LaBerge, "The Paris Health Council, 1802–1848," *BHM*, 49 (1975): 339–352. On public health see also William Coleman, *Death Is a Social Disease: Public Health and Political Economy in Early Industrial France* (Madison: University of Wisconsin Press, 1982).

23. Clément Bressou, *Histoire de la médecine vétérinaire* (Paris: Presses

Universitaires de France, 1970), pp. 76–81; Pierre Huard, "L'instruction médico-chirurgical," in *Enseignement et diffusion des sciences en France au XVIIIe siècle*, ed. René Taton (Paris: Hermann, 1964), pp. 206–209; Huard and Grmek, *Sciences, médecine, pharmacie*, pp. 147–148.

24. Hahn, *Anatomy*, pp. 304–307.

25. Ibid., p. 300.

26. Henri Gouhier, *La jeunesse d'Auguste Comte et la formation du positivisme*, vol. 2, *Saint-Simon jusqu'a à la Restauration* (Paris: J. Vrin, 1936), pp. 31–32, 40–48; Hahn, *Anatomy*, pp. 294–295, 301–304; Triaire, *Récamier*, pp. 94–102.

27. Canguilhem, "Cabanis"; Picavet, *Les idéologues*, pp. 220–224; Staum, *Cabanis*, pp. 109–165, 266–297.

28. Picavet, *Les idéologues*, pp. 206–207, 211, 214.

29. George Rosen, "The Philosophy of Ideology and the Emergence of Modern Medicine in France," *BHM*, 20 (1946): 30–33; Picavet, *Les idéologues*, p. 203; Staum, *Cabanis*, passim.

30. Gouhier, *La jeunesse d'Auguste Comte*, pp. 1–48.

31. Hahn, *Anatomy*, pp. 32–34.

32. Antoine Fourcroy, ed., *La médecine éclairée par les sciences physiques, ou journal des découvertes relatives aux différentes parties de l'art de guérir*, vol. 1 (Paris, 1791), pp. 1–47.

33. Ibid., p. 9.

34. The sixteen fields were: (1) physics, (2) mineralogy, (3) chemistry, (4) botany, both taxonomic and physiological, (5) zoology, (6) human anatomy and animal anatomy compared to that of man, (7) physiology, (8) hygiene, (9) pathology, nosology, and semiology, (10) therapeutics and materia medica, (11) pharmacy, (12) history of epidemic illnesses, those which are endemic and those due to the practice of the arts (occupational medicine), (13) surgery, (14) legal medicine, (15) the veterinary art, (16) the destruction of errors and prejudices in medicine. *La médecine éclairée*, vol. 1 (1791), pp. 15–16.

35. Not surprisingly, Fourcroy gave special emphasis to chemistry. After alluding to the recent studies of respiration and 9digestion, he listed some of the relevant areas of present and future chemical research. See *La médecine éclairée*, vol. 1 (1791), pp. 24–25.

36. Société Royale de Médecine, *Nouveau plan*.

37. For the following, see *Plan général de l'enseignement dans l'Ecole de Santé de Paris: Imprimé par ordre du Comité d'Instruction Publique de la Convention Nationale* (Paris, An III/1795). A note on the final page (p. 49) indicates that the *Plan* was approved by the members of the Comité d'Instruction Publique, Paris, 12 Pluviose An III [2 February 1795]. Signed Fourcroy, (Claude) Prieur, (Jean) Massieu, Plaichard, (Jacques) Bailleul, (Antoine) Thibaudeau, (Luc-François) Lalande, (Louis) Mercier, (Jean) Barailon.

38. Ibid., pp. 16–20.

39. Ibid., pp. 21–22.

3. Bichat's Two Physiologies

Epigraph: Xavier Bichat, *Recherches physiologiques sur la vie et la mort*, 3rd ed. (Paris, An XIII/1805), pp. 322–323.

1. Maurice Genty, "Xavier Bichat, 1771–1802," in *Biographies médicales et scientifiques*, ed. Pierre Huard (Paris: Roger Dacosta, 1972), pp. 181–318, esp. pp. 209–221.

2. Ibid., pp. 185–192.

3. Ibid., pp. 193–196.

4. Genty, "Xavier Bichat," pp. 196–206; "Les lettres de Bichat," *Le progrès médical*, nos. 13–14 (10–24 July 1952), p. 341.

5. *Prospectus de l'école de chirurgie établie au Grand Hospice d'Humanité, ci-devant Hôtel-Dieu de Paris* (Paris, n.d.), pp. 3–8.

6. For discussion of the background of Desault's teaching methods and the circumstances of their development at the Hôtel-Dieu, see Toby Gelfand, *Professionalizing Modern Medicine: Paris Surgeons and Medical Science and Institutions in the 18th Century* (Westport, Conn.: Greenwood Press, 1980), pp. 116–125. For Desault's biography see Pierre Huard and Marie-José Imbault-Huart, "Pierre Desault, 1738–1795," in *Biographies médicales et scientifiques*, ed. Huard, pp. 119–180. The standard contemporary biography of Desault is found in Bichat's two essays, "Notice historique sur la vie de Pierre-Joseph Desault," *Journal de chirurgie*, 4 (1795): 195–217 (my copy has the pagination 1–23), and "Essai sur Desault et sur les progrès que lui doit la chirurgie," *Oeuvres chirurgicales de P.-J. Desault*, vol. 1 (Paris, An VI/1798), pp. 1–48.

7. Toby Gelfand, "The Training of Surgeons in Eighteenth-Century Paris and Its Influence on Medical Education" (diss., Johns Hopkins University, 1973), p. 438.

8. Genty, "Xavier Bichat," pp. 219–226.

9. Ibid., pp. 227–233, 241–242; Huard and Imbault-Huart, "Desault," pp. 122–123.

10. Erwin Ackerknecht, *Medicine at the Paris Hospital, 1794–1848* (Baltimore: Johns Hopkins University Press, 1967), p. 116; Cherest, "Recherches historiques sur la Société: Son origine, ses fondateurs, ses travaux," *Recueil des travaux de la Société Médicale d'Emulation de Paris*, 30 (1850): 5–47; Genty, "Xavier Bichat," pp. 235–237; "Lettres de Bichat," pp. 346–347. Surgeons played a prominent role in the young Société. See Gelfand, *Professionalizing Modern Medicine*, p. 175.

11. On the rise and importance of the private courses see Ackerknecht, *Medicine at the Paris Hospital*, pp. 43–44.

12. Genty, "Xavier Bichat," pp. 239–241. An undated letter of Bichat indicates that he had a regular arrangement for the supply of experimental animals for his courses. "Lettres de Bichat," p. 343 (letter 8).

13. Genty, "Xavier Bichat," pp. 241–242.

14. "Lettres de Bichat," pp. 314–348 (letters 14, 19, 20).

15. Genty, "Xavier Bichat," p. 245; "Lettres de Bichat," p. 345 (letter 8).

16. Institut de France, Académie des Sciences, *Procès verbaux*, vol. 2 (An VIII–IX/1800–1804), pp. 51, 109, 174–175, 401. Hallé, who with Duméril had witnessed the experiments of the *Recherches physiologiques*, gave a favorable report on this work (p. 175).

17. Genty, "Xavier Bichat," pp. 253–256; "Lettres de Bichat," p. 346.

18. Genty, "Xavier Bichat," pp. 257–258; "Lettres de Bichat," pp. 346–347 (letter 17).

19. Ackerknecht, *Medicine at the Paris Hospital*, p. 15.

20. *Mémoires de la Société Médicale d'Emulation*, vol. 1 (Paris, An V/ 1797), Discours préliminaire.

21. Ibid.

22. Bichat papers, MS 46, item 7, Bibliothèque de la Faculté de Médecine, Paris. Since Bichat began to give separate lecture courses in physiology in late 1798, the MS may date from this period. It is noteworthy, however, that the ideas on the interior and exterior lives are much less developed in this MS than they are in the "Mémoire sur les rapports qui existent entre les organes à forme symétrique, et ceux à forme irrégulière," which Bichat published earlier in 1798 in the second volume of the *Mémoires* of the Société Médicale d'Emulation. The *Discours* may therefore date from an earlier course in which some lectures on physiology were to be included. This text has been published by A. Arène, "Essai sur la philosophie de Xavier Bichat," *Archives d'anthropologie criminelle* (1911): 753–825; and in English translation by William R. Albury, "Experiment and Explanation in the Physiology of Bichat and Magendie," *SHB*, 1 (1977): 97–106. For a summary of Bichat's views on experiment as expressed in the *Discours* see Albury, pp. 61–63.

23. *Mémoires de la Société Médicale d'Emulation*, 2 (An VI/1798): 350–370, 371–385, 477–487.

24. Bichat, "Dissertation sur les membranes," pp. 378–379.

25. Bichat, *Recherches physiologiques*, pp. 1–73.

26. Genty, "Xavier Bichat," p. 251; Laignel-Lavastine, "Sources, principes, sillage et critique de l'oeuvre de Bichat," *Bulletin de la Société Française de Philosophie*, 46 (1953): 11–14; Arène, "Xavier Bichat," pp. 779–785; Elizabeth Haigh, "The Roots of the Vitalism of Xavier Bichat," *BHM*, 49 (1975): 72–86; Huard and Imbault-Huart, *Desault*, p. 129; F. Colonna d'Istria, "Cabanis et les origines de la vie psychologique," *Revue du métaphysique et du morale*, 19 (1911): 177–178; François Joseph Picavet, *Les idéologues* (Paris, 1891), pp. 231–233, 435–437; Martin S. Staum, *Cabanis: Enlightenment and Medical Philosophy in the French Revolution* (Princeton: Princeton University Press, 1980), pp. 255–259.

27. Bichat, *Recherches physiologiques*, p. 75.

28. Ibid., pp. 1–2.

29. Ibid., p. 104.

30. Ibid., pp. 79–107.

31. Ibid., pp. 84–89, 127–128.

32. Ibid., pp. 80–84. His view of the continuity between animal and organic properties was connected with his acceptance of the elements of the traditional "great chain of being" complex of ideas, according to which all living forms could be arranged in a single hierarchical, continuous, and complete linear progression. The connection was made explicit in the "Mémoire" of 1798.

33. Bichat was aware of the sequence. See *Anatomie générale appliquée à la physiologie et à la médecine* (Paris, An X/1801), pp. lxxxiii–lxxiv.

34. See, for example, *Oeuvres complètes de Bordeu* (Paris, 1818), pp. 797–847.

35. See, for example, Bichat's remarks on Haller in *Anatomie générale*, p. xxxix. Laignel-Lavastine's suggestion that Bichat obtained his fourfold division of vital properties by applying his theory of the two lives to Haller's distinction of sensibility and irritability contains a part of the truth, but overlooks the generalized character of Bichat's sensibility and contractility, which can have come only from Bordeu. Laignel-Lavastine, "Bichat," p. 12. Cf. Arène, "Xavier Bichat," pp. 777–779.

36. Bichat, "Dissertation sur les membranes," p. 371.

37. Bichat, *Anatomie générale*, pp. lxxix–lxxx.

38. These are listed in Bichat, *Anatomie générale*, pp. lxxix–lxxx. Included in the listing are, of course, the mucous, serous, fibrous, and synovial "membranes" of the *Traité*. Charles Maingault, the editor of an 1818 edition of the *Anatomie générale*, points out that Bichat's tissue classification was soon revised by Dupuytren and Richerand, who reduced the number of tissue systems to eleven. See Xavier Bichat, *Anatomie générale, précédées des recherches physiologiques sur la vie et la mort*, ed. Maingault (Paris, 1818), pp. 35–36, n. 1.

39. Genty, "Xavier Bichat," pp. 279–380.

40. P. Lain Entralgo, "Sensualism and Vitalism in Bichat's *Anatomie Générale*," *JHMAS*, 5 (1948): 60. Bichat was in frequent contact with Pinel. See "Lettres de Bichat," p. 345. On the parallels between Lavoisier's chemistry and Bichat's tissue analysis and the role of Condillac see William Randall Albury, "The Logic of Condillac and the Structure of French Chemical and Biological Theory" (diss., Johns Hopkins University, 1972).

41. Desault included attempts to break down organs into their elementary components in his anatomy lessons. See Marc-Antoine Petit, "Eloge de Desault" (c. 1795), MS 5212, Bibliothèque de la Faculté de Médecine, Paris. On taxonomy, see Paul Delaunay, "La médecine et les idéologues: L. J. Moreau de la Sarthe," *Bulletin de la Société Française d'Histoire de la Médecine*, 14 (1920): 42–43. Bichat explicitly compares his study of the membranes to botanical classification, in "Dissertation sur les membranes," p. 382.

42. On pathological anatomy, see, for example, Bichat, *Anatomie générale*, pp. xcviii–xcix. On materia medica see ibid., pp. xlv–l.

43. Ibid., pp. xxxv–xxxviii.

44. Bichat, *Recherches physiologiques*, pp. 152–154.

45. Ibid., pp. 154–156.

46. Michel Foucault, *Naissance de la clinique: Une archéologie du regard médical* (Paris: Presses Universitaires de France, 1972), pp. 143–149.

47. Bichat was not alone in thus evolving physiological problems out of clinical surgical experience. See Gelfand, "The Training of Surgeons," p. 157; and Huard and Grmek, *La chirurgie moderne: Ses débuts en occident, XVIᵉ–XVIIᵉ–XVIIIᵉ siècles* (Paris: Roger Dacosta, 1968), pp. 15–16.

48. Bichat, *Recherches physiologiques*, pp. 221–224.

49. Ibid.

50. Ibid., pp. 227–228. In a recent work on a twentieth-century physiologist the authors, apparently unaware of these experiments, described crossed circulation as a physiological technique *d'avant garde* c. 1910. Georges de Morsier and Marcel Monnier, *La vie et l'oeuvre de Frédéric Battelli (1867–1941): L'ecole genevoise de physiologie de 1899 à 1941 (J. L. Prevost, F. Battelli, L. Stern)* (Basel and Stuttgart: Schwabe, 1977), p. 80.

51. Bichat, *Recherches physiologiques*, pp. 229–230.

52. Chirurgie. 10° Notes sur les aneurysmes, ff. 27, 28, 29, 30; 13° Notes diverses: Titres, plans, remarques concernant diverse operations, ff. 11, 14, MS 5149, Bibliothèque de la Faculté de Médecine, Paris. Cf. Charles Lenormant, "Bichat chirurgien," *Le progrès médical*, nos. 41–42 (18 October 1941), pp. 729–741; René Leriche, "Bichat chirurgien," *Le progrès médical*, nos. 13–14 (19–24 July 1952), p. 331. A discussion of Desault's work on aneurysms, with descriptions of his operations, may be found in *Oeuvres chirurgicales de P.-J. Desault*, ed. Xavier Bichat, vol. 2 (Paris, An VI/1798), pp. 497–514.

53. In seeking the improvement of surgical technique through animal experiment, Bichat was drawing on and extending a type of investigation that reached back through Desault and John Hunter to the seventeenth century. See Heinrich Buess, *Die historischen Grundlagen der intravenösen Injektion: Ein Beitrag zur Medizingeschichte des 17. Jahrhunderts* (Aarau: H. R. Sauerländer, 1946); Leo M. Zimmerman, "The Evolution of Blood Transfusion," *American Journal of Surgery*, 55 (1942): 613–620; Huard and Grmek, *La chirurgie moderne*, pp. 13–18. Examples are Jean-Louis Petit's series of memoirs on hemostasis, Académie Royale des Sciences, Paris, *Histoire et Mémoires Année 1731*, pp. 85–102; *1732*, pp. 388–397; *1735*, pp. 435–442; *1736*, pp. 244–255; and Jacques Tenon's memoirs on bone exfoliation, ibid., *Année 1758*, pp. 372–418; *1760*, pp. 223–238. The Paris Académie de Chirurgie published reports of experimental esophagotomies and tracheotomies performed by surgeons: Guatani, "Mémoire sur l'oesophagotomie," Académie Royale de Chirurgie, Paris, *Mémoires*, 3 (1757): 351–360; Lescure, "Observation sur une portion d'amande de noyau d'abricot, dans la trachée artère," ibid., 5 (1774): 524–538. On John Hunter see Lloyd Allen Wells, "Aneu-

rysm and Physiologic Surgery," *BHM*, 44 (1970): 411–424. Hunter's discovery of the collateral circulation while carrying out an experiment on the physiological problem of antler growth in a stag, and his subsequent application of the principle in developing a new operation for popliteal aneurysm, are described in John Kobler, *The Reluctant Surgeon: A Biography of John Hunter* (Garden City, N.Y.: Doubleday, 1960), pp. 267–271. In 1783 Hunter was elected to both the Société Royale de Médecine and the Académie Royale de Chirurgie in Paris. See James F. Palmer, ed., *The Works of John Hunter, F.R.S.*, vol. 1 (London, 1835), p. 84; and Hunter's letter to the Académie de Chirurgie, Archives 41 (41), fol. 28, Bibliothèque de l'Académie Nationale de Médecine, Paris.

54. Huard and Grmek, *La chirurgie moderne*, pp. 26–28, 46, 53–55, 102. The difference that a new instrument or technique could make in the hands of a skilled and resourceful surgeon is indicated in Bichat's description of Desault's use of the new sounds. See Bichat, "Essai sur Desault," pp. 27–28.

55. For references to earlier experimental physiology see, for example, Bichat, *Recherches physiologiques*, pp. 205, 245, 261, 272–273. The whole discussion of asphyxiation takes for granted the alteration of the chemical composition of the blood in the lungs; see ibid., pp. 170–173, 190–191.

56. Bichat, *Recherches physiologiques*, pp. 233–234, 258, 308–309, 324. Cuvier's work is cited directly, pp. 332–333.

57. Ibid., pp. 157–159, 314–319.

58. Bichat, *Recherches physiologiques*, pp. 246–247, 272–273, 276, 282; Genty, "Xavier Bichat," p. 252; Institut de France, Académie des Sciences, *Procès verbaux*, 2 (1800–1804): 174–175.

59. Bichat, *Recherches physiologiques*, pp. 322–323. For other citations of clinical-pathological material see pp. 164–165, 169, 173–174, 180–181, 184–190, 231–233, 255, 259, 264–266, 269n, 285n.

60. Ibid., pp. 189, 317.

61. Bichat, "Essai sur Desault," pp. 35–36. In the last line of this passage there is the implicit recognition that surgery is to be seen as a model for a new, more exact medicine. Elsewhere at about the same time Bichat wrote: "Devoted for some time to the study of medicine . . . I now have to consider surgery only as an essential base of all medical knowledge." Quoted in Genty, "Xavier Bichat," p. 242. In his earliest medical studies, before coming to Paris, Desault had been an enthusiastic adherent of Borelli's mechanistic approach to physiology, and had translated and commented on the latter's *De motu animalium*. At some point a reaction had set in, however, and Desault's papers from this period have not been preserved. See Bichat, "Notice historique," pp. 2–3.

62. Arène, "Xavier Bichat."

63. Haigh, "Vitalism of Xavier Bichat," esp. p. 72. Schiller felt that by explaining the mechanism of death by the arrest of organ function Bichat was inconsistent; he should have based the analysis on the loss of properties of the tissues, since these were nominally more fundamental

structures. See Joseph Schiller, *Claude Bernard et les problèmes scientifiques de son temps* (Paris: Editions du Cèdre, 1967), pp. 56–57. The appearance of inconsistency is resolved once the duality of Bichat's physiology is admitted.

64. Lenormant, "Bichat chirurgien"; René Leriche, "Bichat chirurgien."

65. Albury, "Experiment and Explanation."

66. An explicit reference to Foucault appears in ibid., n. 174.

67. Haller's research methods were diverse, and included functional studies in the sense employed here. As Salomon-Bayet has rightly observed, in the eighteenth century natural historical modes of thinking coexisted with other approaches to the study of living things without subordinating them, and the historian must recognize this pluralism. Claire Salomon-Bayet, *L'institution de la science et l'expérience du vivant: Méthode et expérience à l'Académie Royale des Sciences, 1666–1793* (Paris: Flammarion, 1978), pp. 287–288, 333.

68. In a recent study by John V. Pickstone, in contrast, it is just the search for parallels between context and content that leads to effacement of the essential duality of Bichat's physiology. See John V. Pickstone, "Bureaucracy, Liberalism, and the Body in Post-Revolutionary France: Bichat's Physiology and the Paris School of Medicine," *History of Science*, 19 (1981): 115–142.

69. Bichat apparently intended to replace the original Part 1 with a discussion of the alterations of functions in illness, and to this end undertook with Mathieu Salmade experimental inoculations of animals. Part 2 was to be revised, and would include the results of experiments carried out with Nysten on the chemical phenomena of respiration and with Jean Thillaye on galvanism. Genty, "Xavier Bichat," p. 271; Geneviève Genty, *Bichat: médecin du Grand Hospice d'Humanité* (Clermont [Oise]: Thiron, 1943), p. 53.

4. A New Generation and a New Program

Epigraph: John Cross, *Paris et Montpellier, ou tableau de la médecine dans ces deux écoles*, trans. from the English by Elie Revel (Paris and Montpellier, 1820), pp. 44–47. Cross visited France in 1814–1817.

1. *BU*, 35 (Paris, n.d.): 654–657; A. L. Bayle and Auguste Thillaye, *Biographie médicale* (Paris, 1855; rpt. Amsterdam: B. M. Israël, 1967), pp. 913–915.

2. *Nouvelle biographie générale*, 15 (Paris, 1856): 179–181; *Biographies médicales* (Paris, 1929–1931), pp. 133–144. On the chair of physiology at the Paris Faculty of Medicine see A. Corlieu, *Centenaire de la Faculté de Médecine de Paris (1794–1894)* (Paris, 1896), pp. 262–269.

3. Elizabeth B. Gasking, "Serres, Antoine Etienne Reynaud Augustin," *DSB*, 12 (1975): 315–316.

4. J. V. Pickstone, "Vital Actions and Organic Physics: Henri Dutrochet and French Physiology During the 1820s," *BHM*, 50 (1976):

191–212; and Joseph Schiller and Tetty Schiller, *Henri Dutrochet (Henri du Trochet, 1776–1847): Le matérialisme mécaniste et la physiologie générale* (Paris: Albert Blanchard, 1975).

5. Sigalia Dostrovsky, "Savart, Félix," *DSB*, 12 (1975): 129–130; *BU*, 38 (Paris, n.d.): 104–105; Victor A. McKusick and H. Kenneth Wiskind, "Félix Savart (1791–1841), Physician-Physicist: Early Studies Pertinent to the Understanding of Murmurs," *JHMAS*, 14 (1959): 411–423.

6. Vladislav Kruta, "Flourens, Marie-Jean-Pierre," *DSB*, 5 (1972): 44–45.

7. *BU*, 12 (Paris, 1855): 280–282.

8. Pierre Chabert, "Portal, Antoine," *DSB*, 11 (1975): 99–100.

9. *Biographies médicales*, pp. 37–49; "Cours de Physiologie par Chaussier: Ecole de Médecine, 29 Décembre 1806–2 Avril 1807," MS 141, manuscrits du docteur J.-A. Delens (1786–1846), vol. 1: notes de cours, extraits de diverses auteurs, mémoires, etc., pp. 111–212, Bibliothèque de l'Académie Nationale de Médecine, Paris. An example of Chaussier's concentration on anatomical description and nomenclature is his *Exposition sommaire de la structure et des différentes parties de l'encephale ou cerveau; suivant la méthode adoptée à l'Ecole de Médecine de Paris* (Paris, 1807).

10. *BU*, 18 (Paris, 1857): 366–367.

11. *BU*, 11 (Paris, 1855): 511–513.

12. *BU*, 31 (Paris, n.d.): 121–122.

13. P. H. Nysten, *Recherches de physiologie et de chimie pathologiques, pour faire suite à celles de Bichat sur la vie et la mort* (Paris, 1811), pp. xvii–xviii.

14. P. H. Nysten, *Nouvelles expériences galvaniques, faites sur les organes musculaires de l'homme et des animaux à sang rouge* (Paris, An XI/1803).

15. P. H. Nysten, *Recherches de physiologie et de chimie pathologiques.*

16. *Biographies médicales*, pp. 1–28; Léon Delhoume, *Dupuytren* (Paris: J. M. Baillière et fils, 1935).

17. Dupuytren's links with the veterinary school are indicated in a letter from the director of Alfort dated 25 October 1806, in which the director, referring to Dupuytren's "love for science," offered him the post of physician to the school infirmary. Dossier Dupuytren, Académie des Sciences, Paris. See also Delhoume, *Dupuytren*, p. 49.

18. *Biographies médicales*, p. 22. When Dupuytren was elected to the Académie, his most recent original work was the invention of a surgical procedure to correct abnormal openings of the intestine outside the body. In devising the procedure he made use of relevant pieces of anatomical and physiological knowledge and of animal experiment, and invented a new instrument, the *entérotome*. This was not science, however: Dupuytren's career had reversed the path traced by Bichat from experimental surgery to experimental physiology, revealing once again the naturalness of the transition. See the report by Duméril

and Portal on two memoirs of Dupuytren, "Nouvelle méthode pour guérir les anus contre nature," Institut de France, Académie des Sciences, *Procès verbaux*, 8 (1824–1827): 179–181 (meeting of 31 January 1825).

19. Pinel and Hallé, "Rapport fait à la classe de sciences physiques et mathematiques de l'Institut, sur un mémoire de M. Dupuytren, ayant pour titre: Expériences sur l'influence que les nerfs du poumon exercent sur la respiration," *AC*, 63 (1807): 35–48. For Magendie's criticism of these experiments see François Magendie, *Phénomènes physiques de la vie*, vol. 2 (Paris, 1842), pp. 226–229.

20. Pinel and Hallé, "Rapport . . . sur un Mémoire de M. Dupuytren," p. 47.

21. Vladislav Kruta, "Legallois, Julien Jean César," *DSB*, 8 (1973): 132–135; *BU*, 23 (Paris: n.d.): 607–608; Bayle and Thillaye, *Biographie médicale*, pp. 935–936.

22. J. J. C. Legallois, *Expériences sur le principe de la vie, notamment sur celui des mouvemens du coeur, et sur le siège de ce principe* (Paris, 1812); "Section des nerfs de la 8e paire et récurrens" (report of commissioners), Institut de France, Académie des Sciences, *Procès verbaux*, 4 (1808–1811): 335–339; Kruta, "Legallois," p. 134.

23. Percy, Humboldt, and Hallé, "Rapport," Institut de France, Académie des Sciences, *Procès verbaux*, 4 (1808–1811): 522–533.

24. Legallois, *Expériences sur le principe de la vie*, pp. xxiii–xxiv.

25. For Magendie's biography see Claude Bernard, *François Magendie: Leçon d'ouverture du cours de médecine du Collège de France, 29 février 1856* (Paris, 1856); Pierre Flourens, *Eloge historique de François Magendie* (Paris, 1858); J. M. D. Olmsted, *François Magendie, Pioneer in Experimental Physiology and Scientific Medicine in 19th Century France* (New York: Schuman's, 1944); and M. D. Grmek, "Magendie, François," *DSB*, 9 (1974): 6–11.

26. Olmsted, *Magendie*, pp. 14–17; Bernard, *Magendie*, p. 18.

27. Certificates that Magendie issued to three of his Belgian students for attendance in 1813, 1814, and 1815 are preserved in the Magendie papers in the Musée d'Histoire de la Médecine, Faculté de Médecine, Paris.

28. François Magendie, *Formulaire pour la préparation et l'emploi de plusieurs nouveaux médicamens, tels que la noix vomique, la morphine, l'acide prussique, la strychnine, la vératrine, les alcalis des quinquinas, l'iode, etc.* (Paris, 1821).

29. Magendie, "Quelques idées générales sur les phénomènes particuliers aux corps vivans," *Bulletin des Sciences Médicales*, 4 (1809): 145–170. See J. M. D. Olmsted, "Bichat vu par Magendie," *Le progrès médical*, nos. 13–14 (1952), pp. 324–327.

30. A revised second edition of the *Nouveaux élémens* appeared in 1806.

31. The passage under discussion is reprinted in English translation in William Randall Albury, "Experiment and Explanation in the Physi-

ology of Bichat and Magendie," *SHB*, 1 (1977): 47–131, esp. pp. 105–106.

32. Xavier Bichat, *Traité sur les membranes en général, et de diverses membranes en particulier* (Paris, An XI/1802), pp. 190–191. See also T. Bordenave, *Essai sur la physiologie*, vol. 2 (Paris, 1778), p. 43. On the diffusion of similar ideas of organization among students of living things in the decades around 1800 see Joseph Schiller, *La notion d'organisation dans l'histoire de la biologie* (Paris: Maloine, 1978).

33. Michael Gross, "The Lessened Locus of Feelings: A Transformation in French Physiology in the Early Nineteenth Century," *JHB*, 12 (1979): 231–271.

34. Bichat had made statements on the need to concentrate on phenomena accessible to the senses that might have served as direct models for Magendie, for example, "Let us neglect all these idle questions where neither inspection nor experience can guide us. Let us begin to study anatomy where the organs begin to fall under our senses," *Anatomie générale* (Paris, 1801), p. 576.

35. Cf. Claude Bernard's more favorable view of Bichat's vital properties in *Leçons sur les phénomènes de la vie communs aux animaux et aux végétaux* (Paris, 1878), pp. 282–283.

36. For a similar statement by Bichat, see Bichat, *Cours manuscrit*, Bibliothèque de l'Académie Nationale de Médecine, Paris, p. 81. Magendie alludes to parallel ideas expressed by Dupuytren in his teaching at the Hôtel-Dieu. See Magendie, "Quelques idées," p. 168n.

37. François Magendie, *Essai sur les usages du voile du palais, avec quelques propositions sur la fracture du cartilage des côtes*, presented and defended at the Ecole de Médecine, Paris, 24 March 1808 (Paris, 1808).

38. Ibid., pp. 7, 10–11.

5. The Experimentalist in Action

Epigraph: François Magendie, "Mémoire sur le méchanisme de l'absorption chez les animaux à sang rouge et chaud," *JPEP*, 1 (1821): 1–17, esp. p. 6.

1. François Magendie, "Examen de l'action de quelques végétaux sur la moelle épinière," read 24 April 1809 (Paris, 1809).

2. See F. Peron, ed., *Voyage de découvertes aux terres australes* (Paris, 1807–1816). Cuvier's report to the Institute on the expedition is found in vol. 1, pp. i–xv. On the Baudin expedition see also A. L. Jussieu, "Notice sur l'expedition à la Nouvelle-Hollande, entreprise pour des recherches de géographie et d'histoire naturelle," *Annales du Muséum d'Histoire Naturelle*, 5 (1894): 1–11; and John Dunmore, *French Explorers in the Pacific*, vol. 2 (Oxford: Clarendon Press, 1965–1969), pp. 9–40. On Leschenault see *Nouvelle biographie générale*, 29 (Paris, 1859): 923–927. Leschenault reported his findings on *upas tieuté* in "Mémoire sur le *Strychnos tieuté* et l'Antiaris toxicaria, plantes vénéneuses de l'île de

Java, avec le suc desquelles les indigènes empoisonnent leurs flèches; et sur *l'Andira harsfieldii*, plante médicinale du même pays," *Annales du Muséum d'Histoire Naturelle*, 16 (1810): 459–482. In this memoir, published after Magendie's physiological studies, Leschenault, referring to *upas*, noted that "I brought back to Europe a large quantity of it, with which my friend M. Delille, physician and botanist of the expedition to Egypt, and M. Magendie did many interesting experiments that made known the activity and mode of action of these poisons on the animal economy." For further background see M. D. Grmek, *Raisonnement expérimental et recherches toxicologiques chez Claude Bernard* (Geneva: Droz, 1973), pp. 252–261.

3. Magendie, "Examen," pp. 8–9.

4. Ibid., pp. 9–10. Magendie noted that "from more than one hundred fifty experiments that I have made with M. Dupuytren on horses and dogs, it results that the serous membranes absorb every species of liquid, even the most irritating such as bile or concentrated saline solutions, with a surprising promptitude."

5. Ibid., p. 20. Though Magendie and Delille did not say so, their localization of the site of action of the *strychnos* poisons was indebted to recently publicized results of experimental investigations carried out by Legallois. See J. J. C. Legallois, *Expériences sur le principe de la vie*, pp. xviii–xix, 22, 48–49. Magendie later credited Legallois with the experimental demonstration that movements have their seat in the spinal cord. See the *Précis élémentaire*, vol. 1, p. 306.

6. M. P. Earles, "Experiments with Drugs and Poisons in the Seventeenth and Eighteenth Centuries," *Annals of Science*, 19 (1963): 241–254; and idem, "Early Theories of the Mode of Action of Drugs and Poisons," *Annals of Science*, 17 (1961): 97–110.

7. *The Works of John Hunter, F. R. S.,* ed. James F. Palmer (London, 1835), vol. 4, pp. 299–314. On eighteenth-century study of the lymphatics see Nellie B. Eales, "The History of the Lymphatic System, with Special Reference to the Hunter-Monro Controversy," *JHMAS*, 29 (1974): 280–294.

8. Xavier Bichat, *Anatomie générale, appliquée à la physiologie et à la médecine* (Paris, 1801), vol. 2, pp. 594–595.

9. Ibid., pp. 104–106. The experiment was carried out by Dupuytren a few years later, at Alfort, and the results were made available to Magendie.

10. Ibid., pp. 104–106. Despite the hesitation expressed here, Bichat was still teaching within three months of his death that medicines taken into the alimentary canal were absorbed by the lacteals. "Cours de matière médicale, par Bichat, commencé la 4 Prairial de l'an X, cessé le 18 Messidor," MS 1032, p. 27, Bibliothèque de l'Académie Nationale de Médecine, Paris.

11. Reprinted in *JPEP*, 1 (1821): 18–32.

12. Pierre Flandrin, "Expériences sur l'absorption des vaisseaux lymphatiques dans les animaux," *Journal de médecine, chirurgie et phar-*

macie, 85 (1790): 372–382; 87 (1791): 221–238; 90 (1792): 73–88; 92 (1792): 56–80.

13. Magendie, "Mémoire sur les organes de l'absorption," pp. 21–22. Other references by Magendie indicate that he had attended Dupuytren's anatomy course, and had collaborated with him in experiments on absorption at Alfort.

14. Institut de France, Académie des Sciences, *Procès verbaux*, 4 (1808–1811): 208–210 (meeting of 22 May 1809).

15. Ibid., 5 (1812–1815): 142–146 (meeting of 18 January 1813).

16. Magendie, *Précis élémentaire*; the absorption problem is discussed in vol. 2, pp. 166–243, passim.

17. Ibid., vol. 2, pp. 162–163. A similar statement is found on p. 231.

18. Ibid., pp. 229–230.

19. Ibid., p. 221.

20. Ibid., p. 228. For other references to a hydrodynamic approach to the phenomena of the circulatory system, including a quotation from D'Alembert on the difficulties of such an approach, see pp. 250–251, 256–258. For Magendie's views on the relation of physics to physiology see also J. V. Pickstone, "Vital Actions and Organic Physics: Henri Dutrochet and French Physiology During the 1820s," *BHM*, 50 (1976): 191–212.

21. Again, it was Bichat's view that served as foil. *Précis élémentaire*, vol. 2, pp. 216–259.

22. Ibid., pp. 318, 321. Magendie developed these ideas, with a fuller description and critique of contemporary views, in "Mémoire sur l'action des artères dans la circuation," read at the Academy on 17 February 1817, and later published in *JPEP*, 1 (1821), 102–115. The Academy's report on this memoir, signed by Jean Baptiste Biot and Percy, was largely favorable, though it did not definitely commit the Academy to any of Magendie's conclusions. See Institut de France, Académie des sciences, *Procès verbaux*, 7 (1816): 175–179 (meeting of 14 April 1817).

23. "Extrait d'un mémoire sur les vaisseaux lymphatiques des oiseaux, par M. Magendie," Société Philomatique de Paris, *Bulletin des sciences*, ser. 3, vol. 6 (1819), pp. 89–92.

24. Institut de France, Académie des Sciences, *Procès verbaux*, 8 (1824–1827): 155 (meeting of 15 November 1824.)

25. Magendie, "Mémoire sur le méchanisme de l'absorption," pp. 1–17.

26. Ibid., p. 3.

27. Ibid., p. 6.

28. Institut de France, Académie des Sciences, *Procès verbaux*, 7 (1820–1823): 109–111 (meeting of 11 December 1820).

29. Ibid., 5 (1812–1815): 205–208 (meeting of 3 May 1813).

30. Ibid., pp. 205–206.

31. Ibid., pp. 174–179.

32. M. P. Menetrier, "Documents inédits concernant Magendie," *Bulletin de la Société d'Histoire de la Médecine*, 20 (1926): 251–258. In a

letter to Magendie announcing the decree, the Minister of the Interior, Jean Montalivet, remarked that "you owe this mark of favor to the success you have already obtained in the sciences. I do not doubt that you will redouble your efforts to make yourself more and more worthy of it" (pp. 253–254).

33. Among Magendie's papers at the Musée de l'Histoire de la Médecine of the Faculté de Médecine, Paris, is a letter dated 27 March 1816 from Gilbert Chabrol, Conseiller d'Etat and Préfet de la Seine, to the Conseil Général des Hospices recommending Magendie for a place. Chabrol refers to "the witness given to M. Magendie's talents by the Comtes Berthollet and Laplace."

34. On the Montyon prize in experimental physiology see Ernest Maindron, *Les foundations de prix à l'Académie des Sciences: Les lauréats de l'Académie, 1714–1880* (Paris, 1881), pp. 88–92. These pages include a complete list of winners of the prize from 1820 to 1880.

35. Institut de France, Académie des Sciences, *Procès verbaux*, 1 (An IV–VII/1795–1799): 1.

36. Ibid., 2 (An VII–IX/1800–1804): 109 (meeting of 26 Pluviose, An VIII).

37. Ibid., 4 (1808–1811): 478 (meeting of 13 May 1811).

38. Ibid., 7 (1820–1823): 210 (meeting of 30 July 1821).

39. Geoffroy Saint-Hilaire, "1° La zoologie a-t-elle, dans l'Académie des Sciences, une représentation suffisante?—2° La physiologie n'y a-t-elle pas été entièrement oubliée?" *Revue encyclopédique ou analyses et annonces raisonnées des productions les plus remarquables dans la littérature, les sciences et les arts*, 13 (1822): 501–511.

40. Ibid., pp. 509–510.

41. Ibid., pp. 510–511.

42. Ibid., p. 511.

43. Institut de France, Académie des Sciences, *Procès verbaux*, 7 (1820–1823): 246–247 (meetings of 12 November and 19 November 1821).

44. Ibid., 9 (1828–1831): 95 (meeting of 21 July 1828).

45. See Erwin H. Ackerknecht, *Medicine at the Paris Hospital, 1794–1848* (Baltimore: Johns Hopkins University Press, 1967), p. 42.

46. Claude Bernard, *Rapport sur les progrès et la marche de la physiologie générale en France* (Paris, 1867).

47. John Cross, *Paris et Montpellier, ou tableau de la médecine dans ces deux écoles*, trans. from the English by Elie Revel (Paris and Montpellier, 1820), pp. 44–47, 145.

48. Albury, "Experiment and Explanation."

49. Gross, "Lessened Locus of Feelings."

6. Pharmacists and Chemists

Epigraph: "Recherches chimiques sur les quinquina" (first memoir read at the Academy of Science, 11 Septemter 1820), in Pelletier and Caventou, *Analyse chimique des quinquina, suivi d'observations médicales sur l'emploi de la quinine et de la cinchonine* (Paris, 1821), pp. 49–50.

1. Pelletier and Caventou, "Mémoire sur un nouvel alcali végétal (la strychnine) trouvé dans la fève de Saint-Ignace, la noix vomique, etc." (read at the Institute 14 December 1818), *ACP*, 10 (1819): 142–177.

2. For Pelletier's biography see W. A. Smeaton, "Pelletier, Bertrand," *DSB*, 10 (1974): 496–497; Alex Berman, "Pelletier, Pierre-Joseph," ibid., pp. 497–499; and Marcel Delépine, "Joseph Pelletier and Joseph Caventou," *Journal of Chemical Education*, 28 (1951): 454–461.

3. The date of the diploma was 11 April 1810, and not 1809 as Delépine ("Joseph Pelletier and Joseph Caventou," pp. 454, 456) reports. See *Catalogue des maîtres en pharmacie, allant jusqu'à l'année 1790 et continué pour la période 1803–1813*, registre 48, Bibliothèque de la Faculté de Pharmacie, Paris, pp. 211, 212, 213.

4. For Caventou's biography see Marcel Delépine, "Joseph Pelletier and Joseph Caventou," and Alex Berman, "Caventou, Joseph-Bienaimé," *DSB*, 3 (1971): 159–160.

5. Pelletier, "Analyse de l'opoponax," *AC*, 79 (1811): 90–99.

6. Frederic L. Holmes, "Analysis by Fire and Solvent Extractions: The Metamorphosis of a Tradition," *Isis*, 62 (1971): 129–148.

7. For a detailed discussion of Fourcroy's career and work see W. A. Smeaton, *Fourcroy: Chemist and Revolutionary, 1755–1809* (Cambridge: Printed for the author by W. Heffer, 1962).

8. Ibid., pp. 6–10. Bouquet published what may have been the first book devoted specifically to plant analysis, *Introduction à l'étude des corps naturels tirés du règne végétal* (Paris, 1773).

9. Smeaton, *Fourcroy*, p. 138.

10. *HMSRM, 1777–1778* (Paris, 1780), pp. 10–13; and ibid., *1780–1781* (Paris, 1785), pp. 32–37.

11. *Nouveau plan de constitution pour la médecine en France, présenté à l'Assemblée Nationale par la Société Royale de Médecine* (Paris, 1790), pp. 127, 148.

12. Smeaton, *Fourcroy*, pp. 19–20. Some idea of the nature of many of these remedies can be gained from perusing the commissioners' reports. Characteristically, the judgments were negative and the reasons given were the lack of novelty of the remedy and the exaggerated claims made on its behalf. For two examples, signed by Fourcroy, see *Registres contenant le jugement de la Société Royale de Médecine sur les remèdes et les différentes préparations qui lui ont été présentés*, registre 14, Bibliothèque de l'Académie Nationale de Médecine, Paris, pp. 22, 58.

13. *HMSRM, 1776* (Paris, 1779), pp. viii–x.

14. "Mémoire sur l'analyse & les propriétés des différentes parties constituantes de l'ipecacuanha," *HMSRM, 1779* (Paris, 1782), p. 512.

15. Examples may be found in *HMSRM, 1779* (Paris, 1782), pp. 252–263, 520–530; *1780–1781* (Paris, 1785), pp. 238–247, 248–255;

1782–1783 (Paris, 1787), pp. 267–274, 48–55 (Mémoires), 56–65; *1784–1785* (Paris, 1788), pp. 306–318.

16. *HMSRM, 1780–1781* (Paris, 1785), p. 25.

17. *HMSRM, 1782–1783* (Paris, 1787), pp. 341–414, 415–487.

18. *Nouveau plan de constitution pour la médecine en France*, pp. 17, 112–113, 122.

19. Smeaton, *Fourcroy*, pp. 8–9; Fourcroy, *L'art de connoître et d'employer les médicamens dans les maladies qui attaquent le corps humain* (Paris, 1785), pp. xxiii–xxiv.

20. "Mémoire sur la nature des altérations qu'éprouvent quelques humeurs animales, par l'effet des maladies & par l'action des remèdes," *HMSRM, 1782–1783* (Paris, 1787), pp. 488–501; and "Mémoire sur la nature de la fibre charnue ou musculaire & sur le siège de l'irritabilité," ibid., pp. 502–513.

21. Fourcroy, *L'art de connoître*, pp. iv–vi.

22. Ibid., pp. xvii–xix, 68–71.

23. Ibid., pp. xv–xxiv.

24. Smeaton, *Fourcroy*, p. 33.

25. Ibid., p. 73.

26. Antoine Fourcroy, *La médecine éclairée par les sciences physiques, ou journal des découvertes relatives aux différentes parties de l'art de guérir*, vol 1 (Paris, 1791), pp. 29–34.

27. Antoine Fourcroy, *Discours prononcé à la Société libre des pharmaciens le 16 Nivose, an Vme de la République, par le citoyen Fourcroy, lors de son admission dans cette Société* (Paris, An V/1796).

28. Ibid., pp. 32–33.

29. Ibid., pp. 35–36.

30. See the summary of this paper in Smeaton, *Fourcroy*, pp. 163–166.

31. For Vauquelin's biography see Smeaton, *Fourcroy*, pp. 34–35; idem, "Vauquelin, Nicolas Louis," *DSB*, 13 (New York, 1976): 596–598; C. Chatagnon and P. A. Chatagnon, "Les pionniers français de la chimie cérébrale: Les citoyens A.-F. de Fourcroy (1755–1809) et L.-N. Vauquelin (1763–1829)," *Annales médico-psychologiques*, 2 (1954): 14–39, esp. pp. 15–23.

32. Smeaton, *Fourcroy*, pp. 166–176; and Chatagnon and Chatagnon, "Les pionniers," pp. 27–36.

33. C. and P. A. Chatagnon list among the more illustrious of Vauquelin's students Adolphe Bouchardat, Caventou, Robiquet, Chevallier, Payen, Pelletier, Collet-Descotils, Robinet, Gustave Quesneville, Orfila, Friedrich Stromeyer, Thenard, and Chevreul. See "Les pionniers," p. 22. In the report recommending Pelletier's admission to the Société de Pharmacie in 1811 it was noted that Pelletier "has followed M. Vauquelin's private courses in chemistry with the greatest zeal and success for several years." Rapport sur M. Joseph Pelletier, 15 Avril 1811 (Antoine Labarraque, rapporteur), registre 55, pièce 55, Bibliothèque de la Faculté de Pharmacie, Paris.

34. Already in 1789, in an introduction to his teacher Desbois de Rochefort's work on materia medica, the leading clinician Corvisart had complained, in terms similar to Fourcroy's, of polypharmacy and of the poor state of drug therapy generally, and had proposed systematic clinical trials in hospitals as a path to reform. Desbois de Rochefort, *Cours de matière médicale, suivi d'un précis de l'art de formuler*, vol. 1 (Paris, 1789), pp. iii–xl.

35. Bichat, *Anatomie générale*, pp. 9–10.

36. "Cours de matière médicale, par Bichat, commencé le 4 Prairial de l'an X, cessé le 18 Messidor," MS 1032, Bibliothèque de l'Académie Nationale de Médecine, Paris. Bichat's strictures on the older materia medica are found on pp. 1–8.

37. Bayle and Thillaye, *Biographie médicale*, vol. 2 (Paris, 1855; rpt. Amsterdam: B. M. Israël, 1967), p. 871; *BU*, 4 (Paris, 1811): 468; Geneviève Genty, *Bichat: médecin du Grand Hospice d'Humanité* (Clermont [Oise]: Thiron, 1943), p. 54; Maurice Genty, "Xavier Bichat, 1771–1802," in *Biographies médicales et scientifiques*, ed. Pierre Huard (Paris: Roger Dacosta, 1972), pp. 272–273; Erwin H. Ackerknecht, *Medicine at the Paris Hospital, 1794–1848* (Baltimore: Johns Hopkins University Press, 1967), pp. 131–132.

38. C. J. A. Schwilgué, *Traité de matière médicale* (Paris, An XIII/1805), esp. pp. vii–xiv, xliii–xliv; *BU*, 38 (Paris, 1854): 503–504.

39. On Alibert's therapeutics see L. Brodier, *J.-L. Alibert, médecin de l'Hôpital St.-Louis* (Paris: A. Maloine & fils, 1923), pp. 197–218. Alibert's book was *Nouveaux eléments de thérapeutique* (Paris, 1804).

40. J. B. G. Barbier, *Principes généraux de pharmacologie ou de matière médicale: Ouvrage dans lequel on traite de la composition des médicamens et de leurs propriétés actives et curatives* (Paris, 1805), pp. 8, 10. On Barbier see *Nouvelle biographie générale*, 4 (Paris, 1859): 446.

41. On this point and on French pharmacy in the first decades of the nineteenth century see Alex Berman, "The Scientific Tradition in French Hospital Pharmacy," *American Journal of Hospital Pharmacy*, 18 (1961): 110–119; "The Problem of Science in 19th Century French Pharmaceutical Historiography," *Actes du dixième congrès international d'histoire des sciences: Ithaca, 1962* (Paris, 1964), pp. 891–894; and "Conflict and Anomaly in the Scientific Orientation of French Pharmacy, 1800–1873," *BHM*, 37 (1963): 440–462.

42. "Rapport sur l'ouvrage de M. Jadelot médecin, intitulé de l'art d'employer les médicamens, par C. L. Cadet" (undated, but grouped with items from the period 1808–1811), registre 55, Mémoires et Rapports de la Société de Pharmacie, pièce 10, Bibliothèque de la Faculté de Pharmacie, Paris. Jadelot was a physician at the Hôpital des Enfants and the full title of his work was *De l'art d'employer les médicaments, ou du choix des formules dans le traitement des maladies* (Paris, 1805). See Brodier, *J.-L. Alibert*, p. 215n.

43. *Deuxième question du programme des prix proposés par la Société de Pharmacie de Paris, pour l'an 1810.* "Quel est l'état actuel de la phar-

macie en France? Quel est la part qu'elle prend à l'art de guérir & quelles sont les améliorations dont elle est susceptible?" registre 56, pièce 2, Bibliothèque de la Faculté de Pharmacie, Paris.

44. Athenas and Alyon, "Observations sur la thériaque ainsi que sur quelques électuares de ce genre et sur plusieurs teintures et autres préparations pharmaceutiques, par Chansarel, pharmacien à Bordeaux." Meeting of 15 January 1809, "Rapport sur un mémoire de M. Chansarel qui a pour objet de réformer la composition de la thériaque," registre 51, pièces 12, 13, Bibliothèque de la Faculté de Pharmacie, Paris.

45. "Analyse de la réglisse, par M. Robiquet" (Société de Pharmacie, Paris; meeting of 15 August 1809), registre 55, pièce 6, Bibliothèque de la Faculté de Pharmacie, Paris.

46. Réponse à la septième question du programme des prix proposés par la Société Libre de Pharmacie de Paris, pour l'an 1809, "Etablir les propriétés des différentes gommes simples et des gommes résines: En donner une analyse comparée," par M. Chansarel, registre 56, pièce 6, Bibliothèque de la Faculté de Pharmacie, Paris.

47. For a list of the topics of Pelletier's early research see Delépine, "Joseph Pelletier and Joseph Caventou," p. 456.

48. Extrait par M. Robiquet, *ACP*, 4 (1817): 172–185.

49. "Analyse de l'opium: De la morphine et de l'acide méconique, considérés comme parties essentielles de l'opium," traduit par M. Rose, pharmacien à Berlin, *ACP*, 5 (1817): 21–42. For the following see John E. Lesch, "Conceptual Change in an Empirical Science: The Discovery of the First Alkaloids," *Historical Studies in the Physical Sciences*, 11, no. 2 (1981): 305–328.

50. Note appended to Sertürner, "Analyse de l'opium," pp. 41–42.

51. Robiquet, "Observations sur le Mémoire de M. Sertürner, relatif à l'analyse de l'opium," *ACP*, 5 (1817): 275–278. Robiquet's article was followed by a brief study of the "Action de la morphine sur l'économie animale," by Orfila. Ibid., pp. 288–290.

52. Pelletier and Caventou, "Mémoire sur un nouvel alcali végétal (la strychnine)."

53. Ibid., pp. 143–144. Pelletier may have derived this dimension of his approach from association with other members of the Société de Pharmacie. The Société's interest in the connection between botanical taxonomy and chemistry is reflected in one of the prize subjects proposed for 1809. See *Réponse à la troisième question du programme des prix proposés par la Société Libre de Pharmacie de Paris pour l'an 1809*, "Chercher à connoitre les familles naturelles des végétaux par leurs propriétés chimiques, c'est-à-dire déterminer les principes qui les font différer entr'-elles," Par M. Chansarel, pharmacien, Bordeaux; registre 56, pièce 3, Bibliothèque de la Faculté de Pharmacie, Paris.

54. Pelletier and Caventou, "Mémoire sur un nouvel alcali végétal (la strychnine)," pp. 144–145. In a later paper Pelletier and Caventou

indicated that they had begun their work on the *strychnos* plants in 1817 with attempts to analyze *nux vomica*. It was only after failing in this that they turned to St. Ignatius bean and were successful. "Nouvelles recherches sur la strychnine et sur les procédés employées pour son extraction," mémoire lu à l'Académie de Médecine, section de Pharmacie, *Journal de pharmacie et des sciences accessoires*, 8 (1822): 305–306.

55. Pelletier and Caventou, "Mémoire sur un nouvel alcali végétal (la strychnine)," pp. 153–171.

56. Ibid., pp. 167–171.

57. Ibid., pp. 171–177.

58. Ibid., pp. 171–172.

59. "Mémoire sur une nouvelle base salifiable organique trouvée dans la fausse angusture (*Brucaea anti-dysenterica*)," *ACP*, 12 (1819): 113–133; "Examen chimique de plusieurs végétaux de la famille des colchicées, et du principe actif qu'ils renferment: Cévadille (*veratrum sabadilla*); hellebore blanc (*veratrum album*); colchique commun (*colchicum autumnale*)," *ACP*, 14 (1820): 69–83.

60. "Examen chimique des upas," *ACP*, 26 (1824): 44–63.

61. Pelletier and Caventou, "Recherches chimiques sur les quinquina," pp. 49–50. In 1826 the physician Alibert was still refusing to recognize the value of quinine over cinchona, maintaining that only in combination (*en substance*) could cinchona be effective as a medicine. Brodier, *J.-L. Alibert*, p. 206n.

62. "Examen raisonnée des principales préparations pharmaceutiques, ayant le quinquina pour base," in *Analyse chimique des quinquina*, pp. 63–68.

63. "Analyse des observations de M. Double, extrait de la *Revue médicale*, 6ᵉ livraison, 1820," par M. Rouzet, ibid., pp. 74–78. Other physicians carrying out clinical trials on the isolated active principles of cinchona included Pierre Fouquier, Auguste Chomel, Godefroy Coutanceau, Magendie, and Jean Devèze. Ibid., p. 73.

64. A letter by Pelletier and Caventou to the Academy of Science describing details of the manufacture of sulfate of quinine, dated 26 February 1827, was printed in *ACP*, 34 (1827): 331–335. It was reprinted in *Revue du paludisme et de médecine tropicale*, 9 (1951): 41–44. This volume contains a special number devoted to Pelletier and Caventou, which includes several useful articles.

7. Experimental Pharmacology

Epigraph: François Magendie, *Formulaire pour la préparation et l'emploi de plusieurs nouveaux médicamens, tels que la noix vomique, la morphine, l'acide prussique, la strychnine, la vératrine, les alcalis des quinquinas, l'iode, etc.* (Paris, 1821), pp. vi–vii.

1. Magendie was named to the Bureau Central des Hospices Parisiens in 1818, substitute physician at the Salpétrière from 1826 to 1830,

and physician to one of the women's wards of the Hôtel-Dieu in 1830. See J. M. D. Olmsted, *François Magendie* (New York: Schuman's, 1944), pp. 79, 176.

2. The ninth edition of 1836 was a reprinting of the eighth edition of 1834.

3. Erwin H. Ackerknecht, *Medicine at the Paris Hospital, 1794–1848* (Baltimore: Johns Hopkins University Press, 1967), p. 130.

4. See Desbois de Rochefort, *Cours élémentaire de matière médicale, suivi d'un précis de l'art de formuler*, vol. 1 (Paris, 1789), pp. xxiv, 5, 10–11; Bichat, "Cours de matière médicale, par Bichat, commencé le 4 Prairial de l'an X, cessé le 18 Messidor," MS 1032, Bibliothèque de l'Académie Nationale de Médecine, Paris; C. J. A. Schwilgué, *Traité de matière médicale* (Paris, An XIII/1805); L. Brodier, *J.-L. Alibert, médecin de l'Hôpital St.-Louis* (Paris: A. Maloine & fils, 1923), pp. 197–218; and Jean Baptiste Barbier, *Principes généraux de pharmacologie ou de matière médicale: Ouvrage dans lequel on traite de la composition des médicamens et de leurs propriétés actives et curatives* (Paris, 1805).

5. François Magendie, *Leçons sur le choléra-morbus* (Paris, 1832), p. 36.

6. François Magendie, *Phénomènes physiques de la vie*, vol. 4 (Paris, 1842), pp. 218–219.

7. Magendie, *Choléra-morbus*, pp. 238–239.

8. Ibid., pp. 201–202; and Magendie, *Phénomènes physiques*, vol. 2, p. 179. Cf. Pierre Flourens, *Eloge historique de François Magendie* (Paris, 1858), pp. 39–40; and Edwin Lee, *Observations on the Principal Medical Institutions and Practice of France, Italy, and Germany*, 2nd ed. (London, 1843), p. 17.

9. Cuvier, Pinel, Humboldt, and Percy, "On vomiting: Being an Account of a Memoir of M. Magendie read to the Imperial Institute of France on the 1st of March, 1813," *Annals of Philosophy*, 1 (1813): 429–438.

10. Ibid., p. 429.

11. Ibid., p. 436.

12. Ibid., p. 437.

13. Ibid.

14. François Magendie, *De l'influence de l'émétique sur l'homme et les animaux: Mémoire lu à la Première Classe de l'Institut de France, le 23 août 1813, et suivi du rapport fait à la classe par MM. Cuvier, Humboldt, Pinel et Percy* (Paris, 1813).

15. Ibid., pp. 44–62.

16. See Magendie, *Phénomènes physiques*, vol. 4, p. 2.

17. One evidence of Magendie's increasing interest in chemistry is his reference to the work of many recent or current chemical authors, including Lavoisier, Fourcroy, Vauquelin, Berzelius, Thomas Thomson, and Gay-Lussac. Others cited are Armand Séguin, Thenard, Heinrich Vogel, and Humphrey Davy. François Magendie, *Précis élémentaire de physiologie*, vol. 2 (Paris, 1816–1817), pp. 47n, 204, 205, 282–

286, 290, 293–297, 303, 348–351, 354, 363–364, 367–368, 409, 434. For examples of Magendie's references to recent advances in "the science of chemical analysis" and "animal chemistry," see ibid., pp. 94, 242.

18. François Magendie, "Note sure les gaz intestinaux de l'homme sain," *ACP*, 2 (1816): 292–296.

19. "Mémoire sur les propriétés nutritives des substances qui ne contiennent pas d'azote," *ACP*, 3 (1816): 66–77.

20. François Magendie, "Mémoire sur l'emploi de l'acide prussique dans le traitement de plusieurs maladies de poitrine, et particulièrement dans le phthisie pulmonaire" (read at the Academy of Science, 17 November 1817), *ACP*, 6 (1817): 347–360. This paper was reprinted in 1820 with an English translation with preface, translator's introduction, and additions to the text as *Physiological and Chemical Researches on the Use of the Prussic or Hydro-Cyanic Acid in the Treatment of Diseases of the Breast, and Particularly in Phthisis Pulmonalis*, trans. from the French, with notes, by James G. Percival (New Haven, 1820), pp. iii–iv.

21. Magendie, *Physiological and Chemical Researches*, p. 17.

22. For the following, see Percival's introduction to ibid., pp. vi–xv.

23. Gay-Lussac, "Recherches sur l'acide prussique" (presented to the Institute, 18 September 1815), *AC*, 95 (1815): 136–231.

24. Magendie, *Physiological and Chemical Researches*, pp. 21–22.

25. Ibid., p. 28.

26. Ibid., pp. 69–73.

27. Ibid., p. 72.

28. Ackerknecht, *Medicine at the Paris Hospital*, pp. 135–136.

29. Magendie, *Formulaire*, p. 5.

30. Ibid., pp. 6–7.

31. Ibid., pp. 1–12.

32. Ibid., pp. 20–22.

33. Ibid., pp. 52–53.

34. Ibid., pp. 56–76.

35. Ibid., p. v; Magendie, *Formulaire*, 9th ed. (Paris, 1836), p. vi.

36. Letter "A MM. Gay-Lussac et Arago, rédacteurs des *Annales de chimie et de physique*," *ACP*, 3 (1816): 408–410.

37. Magendie, "On the Use of Prussic (Hydrocyanic) Acid in the Treatment of Certain Diseases of the Chest, and Particularly in Phthysis Pulmonalis" (read at the Royal Academy of Sciences, Paris, 17 November 1817, and communicated exclusively to the editor of the Journal of the Royal Institution), *Journal of Science and the Arts*, 4 (1818): 347–358.

38. Magendie, "Recherches physiologiques et médicales sur les causes, les symptomes et le traitement de la gravelle," *ACP*, 7 (1818): 430–436.

39. Ibid., pp. 435–436.

40. *JPEP*, 4 (1824): 1–2; 5 (1825): 252–253, 253n; 7 (1827): 113–121.

41. "Orfila," in E. F. Dubois d'Amiens, *Eloges lus dans les séances*

publiques de l'Académie de Médecine, 1845–1863 (Paris, 1863), pp. 371, 380–381.

42. Magendie, *Physiological and Chemical Researches*, pp. iii–iv.

43. Ibid., p. iv. For other examples of Magendie's attacks on empiricism in the medical community see *JPEP*, 1 (1821): 100; 7 (1827): 67.

44. See, for example, Magendie, *Phénomènes physiques*, vol. 3, pp. 19–30; vol. 4, pp. 2, 3–4.

45. On Broussais see Ackerknecht, *Medicine at the Paris Hospital*, pp. 61–80.

46. See, for example, Magendie, *Choléra-morbus*, pp. 76, 204–206, 258, 259–260. Compare the remarks of Gaspard in *JPEP*, 4 (1824): 1–2.

47. "Feuilleton: Fragmens d'histoire et de biographie médicales contemporaines (M. Magendie)," *Gazette médicale de Paris*, 1 (1830): 223–226, 326–328.

48. *JPEP*, 1 (1821): 65–74, 91–92.

49. *JPEP*, 3 (1823): 377–378. Magendie used *nux vomica*, a *strychnos* poison, in his analysis of the functions of the spinal nerve roots. See *JPEP*, 2 (1822): 367.

50. *JPEP*, 2 (1822): 99–100, 210–224.

51. *JPEP*, 1 (1821): 65–74; 2 (1822): 354–363; 3 (1823): 224–232, 243–246; 266–274; 5 (1825): 319–339; and 7 (1827): 122–126. For other examples see ibid., 3 (1823): 274–300; 7 (1827): 113–121; and 9 (1829): 359–363.

52. Institut de France, Académie des Sciences, *Procès verbaux*, 8 (1824–1827): 498.

53. See, for example, the articles on *nux vomica* and strychnine, *Formulaire* (1836), pp. 1–28. Despite Magendie's optimism, extracts of *nux vomica* varying in strength on a scale of one to five or six were still being sold in Paris as late as 1865. M. D. Grmek, *Raisonnement expérimental et recherches toxicologiques chez Claude Bernard* (Geneva: Droz, 1973), pp. 366–367.

54. *Formulaire* (1836), p. iv. For an example of the new interest see the discussion of the elementary composition of morphine, pp. 49–51. A student of Magendie's, James Blake, made the first systematic attempts to discover connections between the structures of compounds and their physiological action. See William F. Bynum, "Chemical Structure and Pharmacological Action: A Chapter in the History of 19th Century Molecular Pharmacology," *BHM*, 44 (1970): 518–538.

55. *Formulaire* (1836), pp. iii–iv and passim.

56. On the state of physiological chemistry in the 1830s and 1840s see Frederic Lawrence Holmes, *Claude Bernard and Animal Chemistry* (Cambridge, Mass.: Harvard University Press, 1974).

57. Magendie, *Phénomènes physiques*, vol. 1, pp. 21–23, 31, 40–41, 42–43, 79–80, 88–90, 132–133; vol. 2, pp. 177–178, 179–180.

58. Magendie, *Formulaire* (1836), pp. 85–86n.

59. Magendie, *Phénomènes physiques*, vol. 4, pp. 173, 211–218. For other examples of studies of the effects of drugs see ibid., vol. 2, pp. 194–195; and vol. 3, pp. 69–72.

60. Ackerknecht, *Medicine at the Paris Hospital*, p. 158.

61. Magendie, *Choléra-morbus*, pp. 1, 186, 233–234.

62. Ackerknecht suggests that the failure of Broussais's system in the cholera crisis may have done much to undermine his reputation. *Medicine at the Paris Hospital*, p. 67.

63. Magendie, *Choléra-morbus*, p. 36.

64. Ibid., pp. 186, 201–202, 238–239. Cf. Magendie's remarks on René Laënnec's use of emetic in the treatment of rheumatism and pneumonia. *Phénomènes physiques*, vol. 2, pp. 178–179; and Edwin Lee, *Observations on the Principal Medical Institutions and Practice of France, Italy, and Germany* (London, 1843), p. 17.

65. Magendie, *Choléra-morbus*, pp. 36, 204–206, 258.

66. Ibid., p. 12. An example of such an "experiment" was the effect of cholera on the nervous system. See ibid., pp. 151–154.

8. Pathological Physiology

Epigraph: François Magendie, *Leçons sur les fonctions et les maladies du système nerveux*, vol. 1 (Paris, 1839–1841), p. 172.

1. On the interaction of surgery and pathology with experimental physiology from the late seventeenth to the early nineteenth century, see also Max Neuburger, *The Historical Development of Experimental Brain and Spinal Cord Physiology Before Flourens*, trans. and ed. Edwin Clarke (Baltimore: Johns Hopkins University Press, 1981), e.g., pp. 6, 9n10, 11, 19–20, 23–24, 53–54, 66–67, 169–182, 288–289. Magendie's statement is in François Magendie, *Essai sur les usages du voile du palais, avec quelques propositions sur la fracture du cartilage des côtes* (Paris, 1808), p. 6.

2. Erwin H. Ackerknecht, *Medicine at the Paris Hospital, 1794–1848* (Baltimore: Johns Hopkins University Press, 1967), pp. 121–127.

3. See *JPEP*, 1 (1821): 68n; 2 (1822): 99–104, 210–224, 354–363; 3 (1823): 377–378.

4. For evidence of Magendie's continuing involvement in and skill at practical surgery in the 1820s, see *JPEP*, 3 (1823): 385; 7 (1827): 180–202; and Bichat, *Recherches physiologiques sur la vie et la mort*, 4th ed., augmented with notes by Magendie (Paris, 1822), pp. 271–275n.

5. See *JPEP*, 1 (1821): 122, 198–200, 278–279n, 333–340, 337–338n, 387; 3 (1823): 189; 7 (1827): 77–79; 8 (1828): 326–333, 358; and Bichat, *Recherches physiologiques*, pp. 268n, 473–474n, 484–488n.

6. William Coleman, *Georges Cuvier, Zoologist: A Study in the History of Evolution Theory* (Cambridge, Mass.: Harvard University Press, 1964), pp. 81, 89–91. This view of the central importance of the nervous system was later echoed by Desmoulins in his collaborative work with Magendie,

Anatomie des systèmes nerveux des animaux à vertèbres, appliquée à la physiologie et à la zoologie (Paris, 1825), pt. 2, p. 517.

7. For evidence of Cuvier's positive response to the early work of Flourens, see Institut de France, Académie des Sciences, *Procès verbaux*, 7 (1820–1823): 349–354. In 1823 Flourens shared the Academy's Montyon prize in experimental physiology with Magendie's student Fodéra. See ibid., pp. 500, 509. By the early 1820s another former student of Cuvier, Henri Dutrochet, was relating plant movements to the possession by plants of some form of nervous substance. See J. V. Pickstone, "Vital Actions and Organic Physics: Henri Dutrochet and French Physiology During the 1820s," *BHM*, 50 (1976): 191–212.

8. François Joseph Gall and J. C. Spurzheim, *Anatomie et physiologie du système nerveux en général et du cerveau en particulier avec des observations sur la possibilité de reconnaître plusieurs dispositions intellectuelles et morales de l'homme et des animaux par la configuration de leurs têtes*, 4 vols., with 100 engravings (Paris, 1810–1819).

9. Owsei Temkin, "Gall and the Phrenological Movement," *BHM*, 21 (1947): 275–321; Robert M. Young, "The Functions of the Brain: Gall to Ferrier (1806–1886)," *Isis*, 59 (1968): 251–268; idem, "Gall, Franz Joseph," *DSB*, (1972), pp. 250–256; Anthony A. Walsh, "Spurzheim, Johann Christoph," *DSB*, 12 (1975), pp. 596–597. The idea of localization of function in the different parts of the brain was by no means new with Gall and Spurzheim, but had modern antecedents beginning with the work of Thomas Willis (1621–1675). See Neuburger, *Experimental Brain and Spinal Cord Physiology*, pp. 17–109.

10. *JPEP*, 2 (1822): 378–379.

11. See *JPEP*, 2 (1822): 175, 184; 3 (1823): 362–375; 4 (1824): 304; 6 (1826): 50. See also Desmoulins and Magendie, *Anatomie des systèmes nerveux des animaux*, pp. 4–5, 100–101, 109–113, 128–135, 173, 218–219, 222, 224, 226, 229, 232, 238, 240–241, 255, 259, 261, 281, 342.

12. J. M. D. Olmsted, *François Magendie* (New York: Schuman's, 1944), pp. 94–95; *JPEP*, 3 (1823): 234.

13. For evidence that Magendie was open to the principle of cerebral localization, see *JPEP*, 8 (1828): 1–9, 95–96.

14. In 1823 Magendie published the work of an Italian writer who seemed to anticipate some of Flourens's results, and accompanied this paper with some encouraging but condescending remarks aimed at Flourens. *JPEP*, 3 (1823): 95–113.

15. Bichat, *Recherches physiologiques*, pp. 500–501, 507–508. See also Desmoulins, *Anatomie des systèmes nerveux*, pp. 553–558.

16. For evidence of Magendie's access to Paris hospitals prior to 1826 see *JPEP*, 3 (1823): 92n, 114–153, 234, 381, 382; 5 (1825): 33–35; 7 (1827): 67, 69–71.

17. Among physicians who cooperated with Magendie either directly through the sharing of clinical, pathological, or experimental results or as contributors of articles to the *Journal* were Scipion Pinel, Segalas

d'Etchepare, Serres (Pitié), Royer-Collard, Breschet, Caillard (Hôtel-Dieu), Gaspard, Jean Oudet, Koreff, Bérard (Pitié), Pierre Béclard (Pitié), Olivier, Pierre Piorry, and H. Montault. See *JPEP*, 1 (1821): 55–65; 2 (1822): 354–363; 3 (1823): 114–153, 157–161, 232–243, 382; 4 (1824): 1–2, 70–71, 391; 5 (1825): 17–20, 233–265, 340–354; 6 (1826): 45; 7 (1827): 5n; 9 (1829): 53–59, 113–118. This list is far from exhaustive.

18. *JPEP*, 1 (1821): 55–65, 201–204, 384.

19. Ibid., 2 (1822): 99–100. The observation that paralysis caused by brain injuries appeared on the contralateral side of the body from the injury dates from Hippocratic times. See Neuburger, *Experimental Brain and Spinal Cord Physiology*, pp. 53–63.

20. *JPEP*, 2 (1822): 209–210.

21. Ibid., 5 (1825): 340–354; 6 (1826): 45.

22. On pathological anatomy see Ackerknecht, *Medicine at the Paris Hospital*, esp. pp. 164–168.

23. *JPEP*, 2 (1822): 99–100; Bichat, *Recherches physiologiques*, p. 259n. For a more explicit statement of Magendie's views on the separation of physiology and ideology, see Magendie, *Précis élémentaire de physiologie*, vol. 1 (Paris, 1816–1817), p. 171.

24. *JPEP*, 1 (1821): 191. For Magendie's later position see his *Leçons sur les fonctions et les maladies du système nerveux*, vol. 1, p. 3.

25. *JPEP*, 4 (1824): 399.

26. Ibid., 2 (1822): 127–135, 224–231; Institut de France, Académie des Sciences, *Procès verbaux*, 7 (1820–1823): 270, 273, 287, 308.

27. *JPEP*, 4 (1824); 264–266; Desmoulins, *Anatomie des systèmes nerveux*, pp. viii–xx, 534–536.

28. See *JPEP*, 4 (1824): 239–257; Desmoulins, *Anatomie des systèmes nerveux*, pp. 63–64n.

29. Desmoulins, *Anatomie des systèmes nerveux*, pp. xxi–xxii.

30. See Bichat, *Recherches physiologiques*, pp. 260n, 514n.

31. In 1821, 8 out of 43, or about 19 percent of the articles included in Magendie's *Journal* were directly concerned with some aspect of the nervous system. In 1822 the number was 19 of 35, or about 54 percent; and in 1823 it was 12 of 33, or about 36 percent. The relative lengths and qualitative importance of articles are not reflected in these figures.

32. François Magendie, "Expériences sur les fonctions des racines des nerfs rachidiens," *JPEP*, 2 (1822): 276–279.

33. Olmsted, *Magendie*, pp. 93–121; Edwin Clarke and C. D. O'Malley, *The Human Brain and Spinal Cord: A Historical Study Illustrated by Writings from Antiquity to the Twentieth Century* (Berkeley: University of California Press, 1968), pp. 291–309; Paul F. Cranefield, *The Way In and the Way Out: François Magendie, Charles Bell and the Roots of the Spinal Nerves, with a Facsimile of Charles Bell's Annotated Copy of His "Idea of a New Anatomy of the Brain"* (Mount Kisco, N.Y.: Futura, 1974).

34. Clarke and O'Malley, *Human Brain*, pp. 291–296; Neuburger, *Experimental Brain and Spinal Cord Physiology*, pp. 13, 101–109, 247–

258. Neuburger's account provides useful information but is impaired by his overestimation of Bell's role.

35. Charles Bell, *Letters of Sir Charles Bell, Selected from His Correspondence with His Brother George Joseph Bell* (London: John Murray, 1870), pp. 170–171; Clarke and O'Malley, *Human Brain*, p. 297; Cranefield, *The Way In and the Way Out*, item 1811a.

36. Bell, *Letters*, pp. 170–171, 195, 265, 272.

37. Antivivisection sentiments were largely responsible for Bell's failure to pursue his subject experimentally. Even the few experiments he reported he undertook with great reluctance. Bell, *Letters*, pp. 161, 275–276.

38. Charles Bell, "On the Nerves: Giving an Account of Some Experiments on Their Structure and Function, Which Lead to a New Arrangement of the System," *Philosophical Transactions*, 111 (1821): 398–424.

39. Bell, *Letters*, p. 271.

40. Charles Bell, "Recherches anatomiques et physiologiques sur le système nerveux," *JPEP*, 1 (1821): 384–391; Olmsted, *Magendie*, pp. 94–95.

41. Magendie, "Expériences sur les fonctions des racines des nerfs rachidiens."

42. Bell, "Recherches anatomiques." This article was Magendie's summary of what he took to be Bell's ideas.

43. Olmsted, *Magendie*, pp. 95–98.

44. Bell, "Recherches anatomiques," pp. 390–391; Bell, *Letters*, p. 275.

45. François Magendie, "Expériences sur les fonctions des racines des nerfs qui naissent de la moelle épinière," *JPEP*, 2 (1822): 366–371.

46. For the following, see Magendie, "Expériences sur les fonctions des racines des nerfs rachidiens."

47. Ibid.

48. Magendie, "Expériences sur les fonctions des racines des nerfs qui naissent de la moelle épinière," *JPEP*, 3 (1823): 367–368.

49. Magendie, "Sur le siège du mouvement et du sentiment dans la moelle épinière," *JPEP*, 3 (1823): 153–157.

50. Royer-Collard, "Altération de la partie antérieure de la moelle épinière, observée à la maison de santé de Charenton," *JPEP*, 3 (1823): 157–161.

51. Ibid.

52. Magendie, *Système nerveux*, vol. 1, p. 79.

53. *JPEP*, 5 (1825): 27–37. The Italian physician Domenico Cotugno had clearly described the cerebrospinal fluid in 1764. His findings were neglected, however, and Magendie had to call attention to them in 1827. For surveys of work on the subject before Magendie, see David H. M. Wollam, "The Historical Significance of the Cerebrospinal Fluid," *Medical History*, 1 (1957): 91–114; and Clarke and O'Malley, *Human Brain*, pp. 708–733.

54. Ibid., 5 (1825): 31–32; 7 (1827): 9n.
55. Ibid., 1 (1821): 201–204; Bichat, *Recherches physiologiques*, pp. 354–357n.
56. Bichat, *Recherches physiologiques*, pp. 500–501n, 507–508n; *JPEP*, 2 (1822): 276–279.
57. For Magendie's explicit statements to this effect see *JPEP*, 3 (1823): 154–155n, 160–161.
58. Ibid., pp. 173–186, 186–190.
59. Ibid., pp. 232–243.
60. Ibid., 4 (1824): 169; 5 (1825): 27–28, 30–31.
61. Desmoulins, *Anatomie des systèmes nerveux*, p. 44n.
62. *JPEP*, 4 (1824): 391.
63. Ibid., p. 371n; Desmoulins, *Anatomie des systèmes nerveux*, pp. 42–43, 44n, 537–539.
64. *JPEP*, 5 (1824): 27–37.
65. Ibid., pp. 27–37.
66. Ibid., 7 (1827): 1–29, 66–82.
67. Ibid., pp. 6–12.
68. Ibid., pp. 12–18. An analysis by Lassaigne of the cerebrospinal fluid extracted from a horse is included at the end of the memoir, p. 82.
69. Ibid., pp. 18–24.
70. Ibid., pp. 69–77.
71. Ibid., p. 67.
72. Ibid., p. 81. The appeal to veterinarians occurred in the context of Magendie's treatment of a case of immobility in a horse. See pp. 77–79.
73. Xavier Bichat, *Traité des membranes en général, et de diverses membranes en particulier*, new ed., with notes by Magendie (Paris, 1827), pp. vii–ix, 225n, 240n, 247–257nn, 263n. Writing in 1825, Desmoulins had credited Magendie with discovery of the two layers of the arachnoid membrane. Desmoulins, *Anatomie des systèmes nerveux*, p. 121.
74. *JPEP*, 8 (1828): 211–229. In a report of a clinical and pathological observation published the following year, the physician Piorry supported Magendie's findings on the cerebrospinal fluid and at the same time criticized Gall and his followers by name. Ibid., 9 (1829): 53–59.
75. Ibid., 8 (1828): 226–228.
76. Ibid., 7 (1827): 83–84. In the same passage Magendie went on to reflect that "without knowing it, we are under the yoke of the dominant ideas of our century; and regardless of our effort to free ourselves from them, they carry us along or stop us in accordance with their direction and movements."
77. Magendie, *Recherches physiologiques et cliniques sur le liquide céphalo-rachidien ou cérébro-spinal* (Paris, 1842), contains numerous references to clinical and pathological material dating from the period 1830–1834. See esp. pp. 66–124, passim. The importance Magendie attached to the nervous system at this time is also reflected in his lectures on cholera. See his *Leçons sur le choléra-morbus* (Paris, 1832), pp. 259–260.

78. Olmsted, *François Magendie*, pp. 176–177, 187; François Magendie, *Leçons sur les phénomènes physiques de la vie*, 4 vols. (Paris, 1842). For references to Poiseuille's work, see esp. vol. 3, pp. 34–277, passim. Poiseuille's memoir, "Recherches sur la force du coeur aortique," appeared in *JPEP*, 8 (1828): 272–306.

79. Magendie, *Système nerveux*, vol. 2, pp. 2–3, 10, 17–18, 29–30; vol. 2, pp. 5–7. For a reference to recent microscopical results see vol. 1, pp. 142.

80. Ibid., vol. 1, pp. 31–32, 43–46, 53–56, 122–123, 142–148; vol. 2, pp. 11.

81. Ibid., vol. 1, pp. 42–43, 118–124.

82. Ibid., pp. 4–5, 37, 124–125, 172 (quotation), 203–206.

83. Ibid., pp. 49, 59–65, 75–76.

84. See, e.g., ibid., pp. 76–77.

85. Ibid., pp. 70–71.

86. Ibid., pp. 106–107.

87. Ibid., pp. 91–92, 99–103, 192–195, 220–229.

88. See, e.g., Magendie, *Système nerveux*, vol. 1, pp. 73–75, 283–286. Magendie was still practicing surgery in the 1830s. See *Phénomènes physiques*, vol. 4, pp. 353–356.

89. Magendie, *Recherches physiologiques et cliniques*. Magendie's reference to his recent course appears on pp. 40–41. For his remarks on lack of recognition of the fluid, see the preface and pp. 1–4, 8, 35.

90. Ibid., pp. 7–8.

91. Ibid., pp. 66–123.

92. Ibid., pp. 87, 98.

93. Magendie, *Système nerveux*, vol. 1, pp. 203–206; vol. 2, pp. 5–7.

94. Ibid., vol. 2, pp. 2–5. For other remarks to similar effect see Magendie, *Recherches physiologiques et cliniques*, pp. 2, 143.

95. Magendie, "Expériences sur les fonctions des racines des nerfs qui naissent de la moelle épinière," pp. 367–368.

96. François Magendie, "Note sur la sensibilité récurrente," Académie des Sciences, Paris, *Comptes rendus*, 24 (1847): 1130–35; Claude Bernard, *Leçons sur la physiologie et la pathologie du système nerveux*, vol. 1 (Paris, 1858), pp. 25–27, 31, 33.

97. Magendie, *Système nerveux*, vol. 2, pp. 89–102, 341–345.

98. Ibid., pp. 101–102.

99. [François Achilles] Longet, "Fait physiologique relatif aux racines des nerfs rachidiens," Académie des Sciences, Paris, *Comptes rendus*, 8 (1839): 881–883; idem, "Influence des nerfs de la sensibilité sur les nerfs du mouvement," ibid., pp. 919–921.

100. [François Achilles] Longet, "Fonctions des racines des nerfs," Académie des Sciences, Paris, *Comptes rendus*, 11 (1840): 766–767.

101. [François Achilles] Longet, "Recherches sur les propriétés des racines antérieures et des racines postérieures des nerfs spinaux," Académie des Sciences, Paris, *Comptes rendus*, 11 (1840): 1023–25.

102. Olmsted, *Magendie*, p. 215. Members of the commission were, besides Magendie, Henri Ducrotay de Blainville, Pierre Flourens, and Gilbert Breschet. Longet, "Fonctions des racines," pp. 766–767.

103. Magendie, "Sensibilité récurrente," pp. 1130–31.

104. Bernard, *Système nerveux*, vol. 1, pp. 36–61.

105. Magendie, "Sensibilité récurrente."

106. Ibid. See also François Magendie, "De l'influence des nerfs rachidiens sur les mouvements du coeur," Académie des Sciences, Paris, *Comptes rendus*, 25 (1847): 875–879.

107. Magendie, "Sensibilité récurrente."

108. Claude Bernard, "Recherches sur les causes qui peuvent faire varier l'intensité de la sensibilité récurrente," Académie des Sciences, Paris, *Comptes rendus*, 25 (1847): 104–106.

109. Magendie, "De l'influence des nerfs rachidiens"; idem, "Expériences sur l'influence de la sensibilité des nerfs rachidiens sur les mouvements du coeur," Académie des Sciences, Paris, *Comptes rendus*, 25 (1847): 926–928; Bernard, *Système nerveux*, vol. 1, pp. 92–96, 112.

110. Bernard, *Système nerveux*, vol. 1, pp. 54–55.

111. Claude Bernard, *Introduction à l'étude de la médecine expérimentale* (Paris, 1865), pp. 302–313.

112. See Magendie, *Système nerveux*, vol. 2, pp. 56–58, 60–61. Bernard's later presentation on recurrent sensitivity is even more strikingly surgical in character. See Bernard, *Système nerveux*, vol. 1, pp. 56, 61, 62, 63–67, 67–74.

113. Magendie, *Système nerveux*, vol. 2, pp. 88–89; Bernard, *Système nerveux*, vol. 1, pp. 56–88.

9. From Medicine to Biology

Epigraph: Claude Bernard, "Recherches anatomiques et physiologiques sur la corde du tympan, pour servir à l'histoire de l'hemiplégie faciale," *Annales médico-psychologiques*, 1 (1843): 435.

1. J. M. D. Olmsted and E. Harris Olmsted, *Claude Bernard and the Experimental Method in Medicine* (New York: Schuman's, 1952), pp. 28–29; and J. M. D. Olmsted, *François Magendie* (New York: Schuman's, 1944), pp. 210–211.

2. François Magendie, *Leçons sur les fonctions et les maladies du système nerveux*, vol. 2 (Paris, 1839–1841), pp. 223, 224, 233.

3. Olmsted and Olmsted, *Claude Bernard*, pp. 29–30. Note that at least three of those under whom Bernard served before joining Magendie—Velpeau, Maisonneuve, and Manec—were surgeons.

4. Ibid., chs. 4–8; M. D. Grmek, "Bernard, Claude," *DSB*, 2 (1973): 24–34.

5. Grmek, "Bernard," pp. 25–26. Bernard continued to be highly productive into the late 1850s. See Frederic Lawrence Holmes, *Claude Bernard and Animal Chemistry* (Cambridge, Mass.: Harvard University Press, 1974), pp. 441–442.

6. The most important studies are Joseph Schiller, *Claude Bernard et les problèmes scientifiques de son temps* (Paris: Editions du Cèdre, 1967); M. D. Grmek, *Raisonnement expérimental et recherches toxicologiques chez Claude Bernard* (Geneva and Paris: Droz, 1972); Holmes, *Claude Bernard;* and Georges Canguilhem, *Etudes d'histoire et de philosophie des sciences,* 3rd ed. (Paris: Vrin, 1975).

7. Erwin H. Ackerknecht, *Medicine at the Paris Hospital, 1794–1848* (Baltimore: Johns Hopkins University Press, 1967), esp. pp. 121–127.

8. Georges Canguilhem, *Le normal et le pathologique,* 2nd ed. (Paris: Presses Universitaires de France, 1972), pp. 11–67.

9. Georges Canguilhem, "L'idée de médecine expérimentale selon Claude Bernard"; and "Théorie et technique de l'expérimentation chez Claude Bernard," in *Etudes,* pp. 127–142, 143–155. Quote is from "L'idée," p. 140.

10. Holmes, *Claude Bernard,* pp. 443–444.

11. Joseph Schiller, *La notion d'organisation dans l'histoire de la biologie* (Paris: Maloine, 1978), p. 111n.

12. Holmes, *Claude Bernard,* pp. 440–441.

13. Grmek, *Raisonnement expérimentale,* pp. 243, 398–399, 403–405.

14. Marie Jean Pierre Flourens, *Recherches expérimentales sur les propriétés et les fonctions du système nerveux, dans les animaux vertébrés.* 2nd rev. ed. (Paris, 1842). The first edition appeared in 1824.

15. François Achilles Longet, *Anatomie et physiologie du système nerveux de l'homme et des animaux vertébrés,* 2 vols. (Paris, 1842).

16. François Achilles Longet, *Recherches expérimentales et pathologiques sur les propriétés et les fonctions des faisceaux de la moelle épinière et des racines des nerfs rachidiens* (Paris, 1841). Longet states that his clinical and pathological observations were taken mainly from the works of Cruveilhier, Ollivier, Velpeau, Hutin, and John Abercrombie, and from the collections of inaugural theses since 1822 (pp. 86–87). It was no coincidence that 1822 was also the date of Magendie's first demonstration of the distinct functions of the spinal roots.

17. Charles-Edouard Brown, *Recherches et expériences sur la physiologie de la moelle épinière* (Paris, 1846). On Brown-Séquard see also J. M. D. Olmsted, *Charles-Edouard Brown-Séquard: A Nineteenth-Century Neurologist and Endocrinologist* (Baltimore: Johns Hopkins University Press, 1946).

18. *Annales médico-psychologiques,* 1 (1843): introduction, xv–xxvi.

19. Ibid., i–x; P. J. G. Cabanis, *Rapports du physique et du moral de l'homme* (Paris, 1843). Originally published in 1798.

20. *Annales médico-psychologiques,* 1 (1843): xi, xxiii–xxvi.

21. Bernard, "Sur la corde du tympan"; "De l'altération du goût dans la paralysie du nerf facial," *Archives générales de médecine,* 6 (1844): 480–496.

22. Bernard, "Sur la corde du tympan," pp. 408–420.

23. Ibid., p. 425. In a footnote (p. 124) Bernard remarked that the first experiment had been done initially not with the intention of studying the chorda tympani but to determine the influence of the anastomosis of the pneumogastric and the facial nerve. It thus appears that the subject of Bernard's first paper may have been derivative of his own prior work on the pneumogastric and spinal nerves. Those investigations were subsequently published in 1844.

24. Ibid.

25. Bernard, "Sur la corde du tympan," pp. 425–427.

26. Ibid., pp. 427–429. As the Olmsteds have pointed out, Bernard's conclusions were faulty on two counts: the chorda tympani does have a motor effect on salivation, and it does carry sensory as well as motor fibers. In 1857 Bernard corrected himself on the first of these points. Olmsted and Olmsted, *Claude Bernard*, pp. 34–35, 103–105.

27. Bernard, "Sur la corde du tympan," pp. 430–433.

28. Ibid., pp. 435–439.

29. Bernard, "De l'altération du goût."

30. Ibid., pp. 482–485.

31. Ibid., pp. 493–495.

32. Bernard, "Sur les fonctions du nerf spinal," *Archives générales de médecine*, 4 (1844): 397–426; 5 (1844): 51–93.

33. Ibid., 4 (1844): 408–409. Theodor Ludwig Wilhelm Bischoff first presented his results in his doctoral dissertation, "Nervi accessorii Willisii Anatomia et Physiologia" (Heidelberg, 1832). A detailed summary of his argument and experiments appeared as "Ueber die Function des *Nervus accessorius*," *Notizen aus dem Gebiete der Natur- und Heilkunde* (Froriep), 37 (1833): cols. 134–138.

34. Bernard, "Sur les fonctions du nerf spinal," 4 (1844): 412–414.

35. Ibid., 4 (1844): 414–424; 5 (1844): 53–54.

36. Ibid., 5 (1844): 57–59.

37. Ibid., pp. 65–73.

38. Ibid., pp. 78–88.

39. A vivid sense of the common ground of surgery and Bernard's experimental physiology may be gained from the texts and illustrations of two of his works: the survey of laboratory technique in *Leçons de physiologie opératoire* (Paris, 1879), and (with Charles Huette), *Précis iconographique de médecine opératoire et d'anatomie chirurgical*, 2 vols. (Paris, 1854).

40. For example, at one point he referred to "a plethora of pathological cases" showing that the epiglottis by itself did not suffice for closure of the laryngeal opening. Bernard, "Sur les fonctions du nerf spinal," 5 (1844): 75.

41. Ibid., pp. 72–73, 88–89. For other references to pertinent

comparative observations see ibid., 4 (1844): 411, 413 (borrowed from Henri Ducrotay de Blainville), and 5 (1844): 51–53, 84–85.

42. Ibid., 4 (1844): 408–409, 425–426; 5 (1844): 72–73. On this point see also Holmes, *Claude Bernard*, pp. 263–264.

43. Claude Bernard, "Procédé nouveau pour couper la cinquième paire de nerfs dans le crâne," Société de Biologie, Paris, *Comptes rendus et mémoires*, 1 (1849): 104; "Sur les phénomènes réflexes," Société de Biologie, Paris, *Comptes rendus et mémoires*, 4 (1852): 149–151; "Recherches expérimentales sur le grand sympathique et spécialement sur l'influence que la section de ce nerf exerce sur la chaleur animale" (read 7 and 21 December 1853), Société de Biologie, Paris, *Comptes rendus et mémoires*, 5 (1853): 77–107 (*mémoires*, 87–88).

44. Claude Bernard, "Influence de la section des pédoncules cérébelleux moyens sur la composition de l'urine," Société de Biologie, Paris, *Comptes rendus et mémoires*, 1 (1849): 14.

45. Claude Bernard, "L'influence du système nerveux sur la production du sucre dans l'économie animale," Société Philomatique de Paris, *Procès verbaux* (1849): 49–51. See also idem, "Présence du sucre dans les matières vomies par un diabétique," Société de Biologie, Paris, *Comptes rendus et mémoires*, 1 (1849): 4–5.

46. Claude Bernard, "Sur le tournoiement," Société de Biologie, Paris, *Comptes rendus et mémoires*, 1 (1849): 7–9; "Autopsie d'un diabétique," ibid., pp. 80–81; Claude Bernard and Pierre Rayer, "Sur un veau bicéphale," ibid., pp. 7–9. The latter article reflects an awareness of recent German work on the part of both authors. In this case the writer is Friedrich Tiedemann.

47. See, e.g., Bernard, "L'influence du système nerveux sur la production du sucre," and "Sur les phénomènes réflexes." Canguilhem, "Théorie et technique," in *Etudes*, p. 151, has noted another reason for Bernard's use of mammals: because their functions are more differentiated than those of lower animals, those functions are more easily isolable by physiological analysis.

48. Claude Bernard, "Mode d'action de la strychnine sur le système nerveux," Société Philomatique de Paris, *Procès verbaux* (1847): 71–73.

49. Claude Bernard, "Action toxique de l'atropine," Société de Biologie, Paris, *Comptes rendus et mémoires*, 1 (1849): 7–8.

50. Claude Bernard, "Recherches sur les causes qui peuvent faire varier l'intensité de la sensibilité récurrent," Académie des Sciences, Paris, *Comptes rendus*, 25 (1847): 104–106; "Recherches expérimentales sur le grand sympathique," pp. 100–102; "Influence du grand sympathique sur la sensibilité et sur la calorification," Société de Biologie, Paris, *Comptes rendus et mémoires*, 3 (1851): 163–164. The background and significance of Bernard's work on curare and on carbon monoxide poisoning are examined in detail in Grmek, *Raisonnement expérimentale*.

51. Claude Bernard, "Sur la tournoiement qui suit la lésion des pédoncules cérébelleux moyens," Société Philomatique de Paris, *Procès verbaux* (1849): 21–23.

52. Bernard, "Sur le grand sympathique," p. 84. For another example of emphasis on the specification of experimental conditions, see "Sur la sensibilité récurrente et le mouvement réflexe," Société Philomatique de Paris, *Procès verbaux* (1847): 79–81.

53. Bernard, "Sur le grand sympathique," pp. 80–81.

54. As early as 1848, however, Bernard was beginning to consider general questions that would later appear in the *Introduction*, in connection with his plans for a general treatise on physiology. See Holmes, *Claude Bernard*, p. 374.

55. M. D. Grmek, "Robin, Charles Phillipe," *DSB*, 11 (1975): 491–492.

56. Charles Robin, "Sur la direction que se sont proposée en se réunissant les membres fondateurs de la Société de Biologie pour répondre au titre qu'ils ont choisi," Société de Biologie, Paris, *Comptes rendus et mémoires*, 1 (1849): i–xi.

57. See also Holmes, *Claude Bernard*, pp. 401–402.

58. Robin, "Sur la direction," pp. vi–x.

59. Cursory inspection of the first list of members published by the Société shows that all 7 officers, at least 13 of the 15 honorary members, at least 28 of the 32 titular members (the core of the membership), and at least 8 of the 9 national correspondents were either M.D.'s or had some close connection with one of the healing arts. The percentage was apparently lower in the category of associates, many of whom were specialists in basic sciences such as chemistry, botany, zoology, or embryology. The information given for foreign correspondents does not allow an estimate to be made without further research. Société de Biologie, Paris, *Comptes rendus et mémoires*, 2 (1859): v–xiv.

60. Ibid., 1 (1849): 1–2. Rayer, also a member of the Academy of Science, later became dean of the Faculty of Medicine. *Nouvelle biographie générale*, 41 (Paris, 1866), cols. 739–740.

61. For a survey of this journal in its first half century, and further details on the Société de Biologie, see E. Gley, "La Société de Biologie et l'évolution des sciences biologiques en France de 1849 à 1900," in idem, *Essais de philosophie et d'histoire de la biologie* (Paris: Masson, 1900), pp. 168–312. A rough notion of its emphases in the first five years of publication may be gained by comparing the numbers of entries under subject headings listed in the analytical tables of contents. Only the five largest categories for each volume are listed.

Volume/ year	Category	Number of entries (*CR* and *Mém.* combined)
1 (1849)	Physiology	64
	Human and Animal Pathological Anatomy	58

Volume/ year	Category	Number of entries (*CR* and *Mém.* combined)
	Human Pathology	31
	Teratology	23
	Animal Anatomy	21
2 (1850)	Pathology	56
	Animal Physiology	34
	Pathological Anatomy	29
	Teratology	22
	Normal Anatomy	15
3 (1851)	Pathology	24
	Physiology	22
	Pathological Anatomy	13
	Normal Anatomy	12
	Teratology	10
4 (1852)	Pathological Anatomy	45
	Normal Anatomy	16
	Teratology	16
	Physiology	15
	Cysts	11
5 (1853)	Pathological Anatomy	18
	Physiology	15
	Normal Anatomy	11
	Horse	11
	Botany	7

Source: Société de Biologie, Paris, *Comptes rendus et mémoires*, 1 (1849): 161–170; 2 (1850): 227–242; 3 (1851): 267–276; 4 (1852): 487–502; 5 (1853): 325–334.

Index

Absorption, 102–115, 117, 149–151
Académie de Chirurgie (Paris), 41, 74
Académie de Médecine, 127, 191
Academy of Science (Paris), 31, 35, 125, 128, 131, 136, 138, 139, 152, 170, 174, 192–193; and physiology, 9, 17, 24, 26, 42, 83, 115–120, 162; Montyon prize of, 9, 82, 127, 143–144, 162, 173, 198, 210; and experimentalism, 9, 115–116, 168–169, 185; and French Revolution, 36, 41–42; positivism in, 45; Magendie and, 90, 113, 115–118, 120–121, 136, 152, 168–169, 173, 174, 185, 192–193
Ackerknecht, Erwin H., 5, 57–58, 167
Albury, William Randall, 77–79, 122–123
Alfort, *see* Ecole vétérinaire d'Alfort
Alibert, Jean, 55, 134, 147
Alkaloids, 125, 127, 137–144, 158, 163
Alyon, Pierre, 135

Anatomia animata (Haller), 47
Animism, 19
Annales de chimie et de physique, 127, 131, 138
Annales médico-psychologique, 202–203
Auget, Antoine, baron de Montyon, 117
Antivivisection, 10, 260n37
Apothicairerie Générale, 40
Archives générales de médecine, 210
Archiv für Physiologie, 16
Arnold, Friedrich, 206
Autopsy, 6; Claude Bernard and, 6, 195–196, 200, 201, 206, 208, 209, 214, 215, 217, 219; Pierre Desault and, 52–53; Xavier Bichat and, 61; François Magendie and, 107, 136–137, 150, 151–152, 166, 167–168, 170, 171, 179, 180, 181, 182, 183, 188–189, 195–196; and drug action studies, 146, 151–152; Pierre Rayer and, 209, 219

Baglivi, Georgio, 16
Baillarger, Jules, 202

Balzac, Honoré de, 86
Barbier, Jean Baptiste, 134, 147, 164
Barthez, Paul-Joseph, 25–26, 33, 45,
    51, 55, 59, 65–66, 82, 83, 84–85,
    94, 95, 97, 121, 161, 166
Baudelocque, Jean, 39
Baudin expedition, 100
Bell, Charles, 55, 167, 170, 171,
    175–177, 178
Bell-Magendie Law, 175, 211
Bérard, Pierre, 120, 205, 211
Bernard, Claude, 2, 26, 92, 99, 114,
    121, 167, 197–224; collaborates
    with Magendie, 192–196. *See also*
    Introduction à l'étude de la médecine
    expérimentale
Bernoulli, Daniel, 21
Berthollet, Claude, 9, 115, 117–118,
    131
Berzelius, Jöns Jacob, 159
Bichat, Xavier, 4, 24, 26, 30, 43, 44,
    49, 50–79, 84–85, 133, 134, 147,
    161, 162, 166, 167, 182, 183, 188;
    Discours préliminaire, 55, 58;
    Oeuvres chirurgicals de Desault, 56;
    Traité des maladies des voies urinaires
    (Desault, edited by Bichat), 56;
    Traité des membranes, 56–57, 60,
    66, 96, 184; *Anatomie descriptive*,
    57; *Recherches physiologiques sur la*
    vie et la mort, 57, 60, 61–79, 85,
    88, 95, 97, 109, 118; *Anatomie gén-*
    érale, 57, 60, 66, 68, 76, 104, 133,
    134; *Discours sur l'étude de la phy-*
    siologie, 58–60, 62, 74, 76; *Mém-*
    oire, 62; "Dissertation sur les
    membranes," 66; "Essai sur De-
    sault," 75; *Recherches physiologiques*
    sur la vie et la mort (new edition,
    edited by Magendie), 170–171,
    179
Biology, 1–3, 122, 222–224
Bischoff, Theodor, 210–211, 214,
    216, 218
Blumenbach, Johann Friedrich, 21
Boë, Franz de la, 16
Boerhaave, Hermann, 15, 19, 20, 46,
    105, 133
Bordenave, Toussaint, 24, 29, 122
Bordeu, Théophile, 25, 51, 65, 66
Borelli, Giovanni, 16
Botany, 2, 28, 138, 139, 141

Bouillon-Lagrange, Edme, 86
Boyer, Alexis, 86, 89, 97
Boyle, Robert, 16
Braconnot, Henri, 139
Breschet, Gilbert, 180
Brodie, Benjamin, 159
Bromine, 163
Broussais, François, 43, 160–161,
    164, 165
Broussonnet, Pierre, 28
Brown, Charles-Edward (later Brown-
    Séquard), 201–202
Brucine, 127, 142, 162
Bucquet, Jean Baptiste, 128, 129
Burdach, Karl, 187
Burdin, Claude, 55
Bureau des Longitudes, 36

Cabanis, Pierre, 26, 32–37, 44–45,
    48, 55, 60, 202. *See also Du degré de*
    certitude de la médecine
Cadet, Charles Louis, 134–135
Caffeine, 127
Caldani, Leopoldo, 21, 22
Canguilhem, George, 199
Cartier, Louis-Vincent, 52
Carus, Julius, 187
Caventou, Joseph, 91, 124, 125, 127,
    132, 133, 136, 137, 138–144, 156
Cerebrospinal fluid, 179–186, 188–
    191
Cerise, Laurent, 202
Chaptal, Jean, 57
Charité Hospital, 53, 155, 197, 209,
    219
Chaussier, François, 37–39, 82, 83,
    84–85, 86, 90, 94, 120, 161
Chef des Travaux Anatomiques, 41
Chemistry, 2, 17, 28, 125–144, 146–
    147, 152–153, 155–164, 183,
    200, 236n35, 254n17, 256n54
Chevreul, Michel, 146–147, 152,
    183
Cholera, 148, 164–165
Chomel, Auguste, 209
Chopart, François, 54
Chorda tympani nerve, 203–210,
    222
Cinchona bark, 133, 143
Cinchonine, 127, 142, 158, 164
Classification, 122, 252n53; of tis-
    sues, 13, 20, 83, 239n38; in Xavier

Bichat's vital properties doctrine and tissue anatomy, 60, 61, 63–64, 66–68, 76–77, 239n38; and materia medica, 134, 147, 148, 158; and Georges Cuvier, 169; and Charles Bell, 176
Codeine, 163, 164
Coleman, William, 21
Collège de Chirurgie (Paris), 15, 24, 27, 29, 38, 39, 41
Collège de Chirurgie (Montpellier), 24–25
Collège de France, 54, 83, 91, 92, 121, 164, 165, 186, 191, 195, 197, 198, 201
Collège de Pharmacie, 28, 39–40
Collège des Etudiants en Médecine, 86
Collège Royal, 26, 31
Comité de Santé, 34–35, 46
Commission Administrative des Hospices, 57
Comparative anatomy: and Georges Cuvier, 5, 112–113, 167, 168–170, 173–174, 186; and Albrecht Haller, 20; and Karl Friedrich Kielmeyer, 21; and School of Montpellier, 25; at Ecole vétérinaire d'Alfort, 28; and Antoine Fourcroy, 46; and André Duméril, 82; and Etienne Serres, 82; and François Magendie, 112–113, 116, 167, 168–170, 173–175, 186, 187, 196; and John Hunter, 112; and Antoine Desmoulins, 169–170, 173–174, 186; and Claude Bernard, 196, 203, 205, 207, 208, 217–218, 219–220, 222, 223.
Comte, Auguste, 45, 222, 223
Condillac, Etienne, 33, 45
Conseil de Salubrité, 40
Conseil d'Hygiène Publique et de Salubrité, 40
Conservatoire des Arts et Métiers, 36
Cornette, Claude, 129, 130
Corvisart des Marets, Jean, 53, 54, 55, 95
Cotugno, Domenico, 21
Couerbe, J. P., 163
Cranioscopy, 169–170, 185, 187
Cross, John, 80, 121, 159

Cruikshank, William, 22–23
Cruveilhier, Jean, 193
Cullen, William, 22–23, 103, 133
Curare, 198, 200
Cuvier, Georges, 5, 57, 73, 82, 88, 90, 92, 95–96, 110, 112–113, 122, 123, 167, 168–170, 173, 174, 186

Daubenton, Louis, 28, 57
Delille, Rafenau, 98, 106, 139, 140, 141, 149, 160
Desault, Pierre, 34, 39, 50, 52–54, 56, 57, 58, 59, 60–61, 69, 72, 73, 74, 75, 76, 89, 95
Descartes, René, 15
Desgenettes, René, 120
Desmoulins, Antoine, 169, 173, 174, 181, 186
Desportes, Henri, 139
Diemerbroek, Isbrand de, 74
Dionis, Pierre, 26
Dolman, Claude, 21
Double, François, 143
Drugs, 6, 91, 102–103, 114, 117, 124, 125–144, 145–165; and the nervous system, 168; and Claude Bernard, 200, 209, 220, 251n34
Dubois, Antoine, 39
Du degré de certitude de la médecine, 32–34, 44–45. *See also* Cabanis, Pierre
Dumas, Pierre, 82, 83, 84–85, 94
Duméril, André, 55, 56, 58, 74, 81–82, 84–85, 97, 113, 118–119, 120, 169
Dupuy, Jean, 86, 87, 168, 176
Dupuytren, Guillaume, 5, 39, 43, 55, 56, 81, 82, 84–87, 89, 90, 106, 110, 122, 123
Dutrochet, Henri, 81, 82, 84–85

Ecole Centrale des Travaux Publiques, *see* Ecole Polytechnique
Ecole de Médecine (Paris), 82, 86, 89, 97
Ecole de Santé (Paris), 32, 34, 36, 39, 47, 54, 55, 56, 57, 58, 83, 84, 87
Ecole Gratuite de Pharmacie, 40
Ecole Normale, 36
Ecole Polytechnique, 36, 42

Ecole Pratique, 6, 39, 84, 90, 201, 223
Ecoles Centrales, 36
Ecole Supérieure de Pharmacie, 124, 126–127, 132, 133
Ecole vétérinaire d'Alfort, 7, 28, 40–41, 86, 87, 105, 109–110, 168, 170, 176, 185–186
Edwards, William, 82, 84–85, 119
Ehrenberg, Christian, 187
Eller, Johann Theodor, 21
Embryology, 1, 2
Emetics, 153, 156, 157, 164
Emetine, 127, 136–137, 146, 150–151, 152, 158
Empiricism, 5, 45, 48, 58–60, 159, 184, 199, 210, 218, 220–221
Enlightenment, 43, 44
Erasistratus, 100
Esquirol, Jean, 43, 202
Evolution, 1, 2
Experimental determinism, 2, 192, 220
Experiment or Experimentalism, 1, 18–19, 20, 21, 22, 23, 24, 25, 34, 35, 46, 47–48, 61–62, 115, 116–124, 125–126, 140–142; in Bichat, 68–79; new generation, 81–89; Magendie 100–110; in experimental pharmacology, 145–165; in pathological physiology, 166–196; in Claude Bernard, 197–224

Faculty of Medicine (Paris), 5, 28, 32–33, 36, 38, 39, 41, 43, 82, 86, 90, 120, 155, 179, 180, 192, 193, 222
Falret, Jean, 197
Fernel, Jean, 15
First Class of the Institute, 41–42, 85; Montyon prize of, 9; and physiology, 9, 48, 81, 115–120, 123–124, 145; healing arts in, 9, 121; Xavier Bichat and, 56–57, 74; Jean Hallé and, 83; Guillaume Dupuytren and, 87; Julien Legallois and, 88–89; François Magendie and, 90–91, 92, 94, 95, 99–100, 110, 117–118, 149–150, 151, 152; and experimentalism, 123–124. See also Academy of Science (Paris)
Flandrin, Pierre, 7, 41, 105–106, 109

Flourens, Pierre, 82, 84–85, 121, 123, 169, 170, 174, 201
Fontana, Felice, 18, 21, 22, 24
Fontenelle, Bernard de, 45
Foucault, Michel, 69, 77
Fourcroy, Antoine, 28, 35, 37–38, 42, 45–47, 83, 128–129, 130–132, 133, 144; La Médecine éclairée par les sciences physiques, 42, 45–46, 58, 131; L'art de connaître et d'employer les médicamens dans les maladies qui attaquent le corps humain, 130–132
Fragonard, Honoré, 41, 56

Galen, 14, 100, 175
Galenic teaching, 12, 16–17, 19
Gall, Franz Joseph, 167, 169–170, 185, 187
Galvani, Luigi, 21, 55
Gaspard, Bernard, 159, 162
Gassicourt, Charles de, 40
Gay-Lussac, Joseph, 5, 115, 138, 142, 153, 154, 164
Gazette médicale de Paris, 161
Gerdy, Pierre-Nicolas, 193
German universities, 2, 3
Gilbert, Ludwig Wilhelm, 137–138
Girard, Jean, 41
Girtanner, Cristoph, 21
Goodwyn, Edmund, 55
Gouhier, Henri, 45
Granville, Augustus, 159
Great Windmill Street School of Anatomy, 22
Grmek, Mirko, 198, 200
Gross, Michael, 123
Guarini, Luigi, 206, 207–208
Guéneau de Mussy, François, 209

Hales, Stephen, 18, 22–23, 24
Hall, Marshall, 202
Hallé, Jean, 56, 74, 82, 83, 84–85, 87, 88
Haller, Albrecht, 12, 15, 20–21, 22, 46, 47, 55, 59, 65, 69, 73, 78, 81, 88, 100, 105, 119
Harvey, William, 12, 17, 51, 73, 76, 100, 122, 176
Haüy, René-Just, 126
Helmont, Jan van, 16
Herophilus, 100, 175

Hewson, G., 112–113
Hewson, William, 22–23
Hippocratic writings, 12, 33, 44, 45, 199
Hoffmann, Friedrich, 19, 21
Holism, 9, 96; in programs for materia medica, 147
Holmes, Frederic, 198, 200
Home, Everard, 159
Hooke, Robert, 73
Hôpital de Perfectionnement, 39
Hôpital Saint-Louis, 134
Hospice des Enfants-Trouvés, 85, 170, 180
Hospitals: and Paris clinical school, 5, 54, 57–58, 167–168; and pharmacy, 40; and medical education, 51–54, 57–58; and Pierre Nysten, 85; and Guillaume Dupuytren, 86; and Julien Legallois, 87, 88; and François Magendie, 117, 146, 165, 167–168, 171, 174, 182, 185, 188; and experimental pharmacology, 145, 146, 157; and pathological physiology, 165, 167–168, 171, 174, 182, 185, 188; and Claude Bernard, 209, 210; and Société de Biologie, 222
Hôtel-Dieu (Grand Hospice d'Humanité), 34, 50, 51, 52–54, 55, 56, 57, 61, 69, 74, 75, 86, 90, 92, 133, 162, 164, 182, 183, 186, 197, 205, 209
Humboldt, Alexander von, 18, 24, 74, 88
Hunter, John, 22–23, 41, 55, 72, 73, 103, 105, 109, 112
Hunter, William, 22, 103, 105
Husson, Henri, 55
Huzard, Jean Baptiste, 57

Iatrochemistry, 15, 16–17
Iatromechanism, 16–17, 19, 25, 48
Iatrophysics, 15
Ignatia amara, 102
Institutionalization, 118–121
Internal medicine, 5, 6, 75–76
Introduction à l'étude de la médecine expérimentale, 2, 5, 167, 191, 199–200; *See also* Bernard, Claude
Iodine, 158

Ipecacuanha, 127, 129, 136, 153, 156
Istituto delle Scienze (Bologna), 22

Jardin des Apothicaires, 41
Jardin des Plantes (Montpellier), 25
Jardin du Roi, 26, 31, 41, 83
Jurin, James, 22
Jussieu, Antoine Laurent de, 57, 110

Kant, Immanuel, 9
Kielmeyer, Carl Friedrich, 21

Labillardière, Jacques, 110
Lafargue, G. V., 220
Laplace, Pierre Simon, 4–5, 90, 117–118
Lassaigne, Jean, 183
Lassone, Joseph, 129, 130
Latreille, Pierre, 118–119
Lauth, Thomas, 113
Lavoisier, Antoine, 73, 126, 131
Leclerc, Claude, 86
Legallois, Julien, 6, 81, 84–85, 87–89, 122, 123, 149–150, 151, 168, 169, 171, 175, 179
Lerminier, Théodoric, 155
Leschenault, Louis, 100, 142
Lieberkühn, Johannes Nathanael, 21
Lieutaud, Joseph, 74
Linnaeus, Carolus, 23, 138
Locke, John, 33
Longet, Achilles, 192–193, 194, 197, 201, 202, 211, 214, 220
Lorry, Anne Charles, 24
Louis, Antoine, 24, 29, 34–35
Lower, Richard, 16
Ludwig, Carl, 2
Lymphatic system, 103–113

Macartney, James, 159
Magendie, François, 4, 5, 26, 30, 77, 81, 84–85, 99–118, 120–124, 145, 166; *Formulaire pour la préparation et l'emploi de plusieurs nouveaux médicamens*, 8, 146, 157–158, 159, 162–164, 186; *Journal de physiologie expérimentale et pathologique*, 8, 91, 92, 166, 169, 171–173, 174, 179, 180, 181, 186, 223; career, 89–92; *Précis élémentaire de physiologie*, 90, 110–112, 113, 124,

Magendie, François, *(Continued)*
152; programmatic position, 92–98;
on absorption, 104–110; and Joseph Caventou and Joseph Pelletier, 136–144; on ipecacuanha,
136–137; and Rafenau Delille,
140–141; and experimental pharmacology, 146–165; and pathological physiology, 166–196; and
Claude Bernard, 197, 198, 199,
200, 201, 202, 205, 208, 210–
211, 218–219, 220, 221–222, 223–
224
Maisonneuve, Jules, 197
Malpighi, Marcello, 16, 17
Mascagni, Paolo, 55
Materia medica, 6, 126, 129–135,
138, 143, 144, 147–148, 156–
157, 251n34
Meconine, 163
Medical education, 31, 35–39, 43,
47–48, 51–54, 57–58, 80–84,
89–90, 109, 121, 210
Medical profession, 32–33, 37
Medical University (Montpellier),
24–25
Meteorology, 42
Microscopy, 5, 19
Mirbel, Charles, 110
Molière, 33
Montaigne, M. E. de, 33
Montpellier, School of, 19, 24–26,
51, 52, 59, 64, 65–66, 69, 76, 81,
83, 121, 166
Montyon Prize (Academy of Sci-
ence), 9, 127, 143, 162, 173, 193,
210
Morgagni, Giovanni, 74
Morphine, 137–138, 140, 158
Morphology, 9
Morveau, Guyton de, 131
Müller, Johannes, 9, 211
Muséum Nationale d'Histoire Natu-
relle, 42, 48, 82, 121, 142, 167

Narceine, 163
Narcotine, 158
Natural theology, 10
Naturphilosophie, 9, 21, 187
Nervous system: and pathological
physiology, 166–196; Claude Ber-
nard on, 201–222

Newton, Isaac, 15, 16, 25, 33, 68,
79
Newtonianism, 25, 26, 59–60, 65–
66, 68, 76, 79, 95
Nosology, 23, 25, 199
Nux vomica, 102, 139, 140, 141, 142,
153, 158, 162, 164
Nysten, Pierre, 6, 81, 82, 84–85, 89,
118, 122, 123, 162

Oken, Lorenz, 187
Ollivier, Charles, 184
Opium, 158
Opoponax, 127–128
Orfila, Matthieu, 43, 120, 160
Oxford School, 16, 17

Paduan school, 14
Paracelsus, 16
Paris clinical school, 5, 54, 167–168,
199–201, 223
Pathological anatomy, 5, 54, 61, 68,
86, 146, 152, 165, 168, 170, 171,
172, 179–191, 199, 202–203,
222, 223
Pathological physiology, 7, 85, 166–
196, 219, 223
Pathology, 8; and Pierre Desault, 61;
and Xavier Bichat, 61, 66, 74–75,
85; and Philippe Pinel, 66, 74–75;
and Antoine Portal, 83; and Pierre
Nysten, 85; and François Magendie, 92, 105, 115, 166, 171–173,
174–175, 179–191, 196, 221; and
Pierre Flandrin, 105; and Jean Alibert, 134; and Claude Bernard,
196, 199, 201, 207, 208–210, 217,
218–219, 221; and *Annales médico-
psychologiques*, 202–203; and Société de Biologie, 222–224
Pelletan, Philippe, 110
Pelletier, Joseph, 90, 124, 125, 126–
128, 132, 133, 136–137, 138–
147, 153, 156, 157, 163
Pelouze, Théophile, 163
Percy, Pierre-François, 12, 88, 116–
117
Petit, Jean-Louis, 24
Petit, Marc-Antoine, 51, 52
Pharmacie Centrale des Hospices,
40
Pharmacology, 7, 91, 134, 145–165

Philip, Wilson, 159
Phrenology, 169–170, 185, 187
Pinel, Philippe, 12, 34, 44, 55, 61, 66,
  87, 90, 110, 116–117, 118, 133,
  147, 169, 172, 202
Pinel, Scipion, 171, 219
Pitié Hospital, 160
Portal, Antoine, 53–54, 74, 82, 83,
  84–85, 110
Positivism, 45, 48, 222–223
Pourfour du Petit, François, 24
Principe vital, 25–26
Private or free courses, 7, 29, 51,
  55–56
Prochaska, Georgius, 21
Profession, 3, 17, 26–29, 81, 125–
  126, 134–136, 145, 159
Professionalization, 26–29, 51, 81,
  121, 125–126, 134–136
Programmatic statements: Bichat,
  58–60, 68, 76–77; Magendie, 92–
  98; Pelletier and Caventou, 138–
  139; pathological physiology, 166–
  168, 171–175; Claude Bernard,
  199–200, 210, 218, 220–221
Prussic acid, 146, 153–156, 158,
  162, 164
Prussic ether, 163, 164
Purkinje, Jan, 187

Quinine, 127, 142, 143, 144, 158,
  162, 164

Rayer, Pierre, 197, 209, 219, 222–
  223
Réaumur, René Antoine de, 18, 24
Récamier, Joseph, 91
Recurrent sensitivity, 191–196, 220
Reil, Johann Christian, 16, 21
Remak, Robert, 187
Rey, Jean-Vincent, 52
Richerand, Anthelme, 44, 81, 84–85,
  94, 97
Robin, Charles-Phillipe, 222–223
Robiquet, Pierre, 127, 135, 138,
  142, 162, 163
Rochefort, Desbois de, 133, 147
Rouelle, Guillaume, 128
Rousseau, Jean-Jacques, 33
Roussel, Pierre, 44
Royal College of Physicians (London), 16

Royal Society of London, 23, 176
Royer-Collard, Antoine, 179
Rullier, P., 179
Ruysch, Frederik, 105

Sabatier, Raphael, 110
Saint-Hilaire, Etienne Geoffroy, 12,
  86, 118–120, 174
St. Ignatius bean, 139, 140, 141, 142
Salpétrière Hospital, 133, 182, 185,
  197, 202
Saucerotte, Nicholas, 24
Sauvages, François Boissier de, 25,
  65–66
Savart, Félix, 82, 84–85
Schiller, Joseph, 200
Schwilgué, C. J. A., 133–134, 147
Segalas d'Etchepare, Pierre, 162
Serres, Etienne, 81, 82, 84–85, 160
Sertürner, Friedrich Wilhelm, 137–
  138
Shaw, John, 170, 176, 177
Société de Biologie (Paris), 8, 168,
  201, 219, 222–224, 267nn59,
  61
Société de l'Ecole de Médecine de
  Paris, 57
Société de Pharmacie, 126, 134,
  135
Société des Pharmaciens de Paris,
  131, 133
Société Libre des Pharmaciens de
  Paris, 40
Société Médicale d'Emulation, 44,
  50, 55, 58, 60, 83, 86
Société Royale de Médecine, 6, 24,
  28, 32, 34, 40, 46, 83, 126, 128,
  129–130, 131, 132, 235n21
Société Royale des Sciences (Mont-
  pellier), 25
Soemmerring, Samuel, 55
Sorbonne, 121, 198
Spallanzani, Lazzaro, 14, 18, 21, 22,
  24, 51, 55, 59, 73, 82
Spinal nerve roots, 175–179, 191–
  196, 201, 210–211, 216
Spinal or accessory (eleventh cranial)
  nerve, 210–218, 222
Spurzheim, Johann, 167, 169–170,
  176, 180, 185, 187
Stahl, Georg Ernst, 19, 21, 25, 64–
  65

Steno, Nicholas, 73
Strychnine, 140, 141, 142, 143, 158, 162
Strychnos plants, 98, 102, 104, 106–108, 125, 127, 138–139, 141, 142, 146, 148, 150, 151, 155, 156, 160
Stuart, Alexander, 22–23
Surgery, 17, 18–19, 20, 22, 24, 47; and internal medicine, 5, 46; professionalization of, 27; training in, 27, 52–54; and Xavier Bichat, 50, 51–54, 60, 61, 69–77; and Pierre Desault, 50, 52–54, 56, 57, 60; and Guillaume Dupuytren, 86–87, 243n18; and François Magendie, 90, 99, 109, 168, 175, 177–178, 189, 221; and pathological physiology, 166; and spinal nerve roots, 175, 177–178; and Claude Bernard, 200–201, 208, 217, 218, 221
Swammerdam, Jan, 105
Sylvius, *see* Boë, Franz de la

Teleomechanism, 9
Tenon, Jacques, 24, 31, 35
Thenard, Louis, 87, 115, 162
Therapeutic skepticism, 5, 145–148, 156–157
Therapeutics, 68, 115, 126, 133–135, 138–139, 142–143, 145–149, 150, 154–165, 200
Thouret, Michel, 55
Tingry, Pierre, 130

Trembley, Abraham, 18, 24
Treviranus, Gottfried, 187

University of Bologna, 22
University of Edinburgh, 22
University of Glasgow, 22
University of Göttingen, 20
University of Halle, 19
University of Leiden, 19, 20
Upas tieuté, 98, 100–102, 106–108, 141, 142, 149, 153

Valentin, Gabriel, 187
Vauquelin, Nicolas, 57, 86, 132–133, 135
Velpeau, Alfred, 197, 209
Veratrine, 127, 142, 158, 162
Vesalius, Andreas, 12
Veterinary medicine, 7, 28, 105. *See also* Ecole vétérinaire d'Alfort
Vicq d'Azyr, Féliz, 28, 34, 41
Villermé, Louis, 162
Vital force, 92–96
Vitalism, 26, 51, 232n36
Vital properties, 13, 62–68, 69, 76–79, 80, 87, 93–97, 111, 123, 239nn32,35
Vivisection, 9
Vomiting, research on, 117, 149–152
Voyages of discovery, 81, 100

Whytt, Robert, 22–23, 175
Wollaston, William, 159

Zoology, 2, 118–121, 123, 173–174